THE SECRET WAR BETWEEN DOWNLOADING AND UPLOADING

PETER LUNENFELD

THE SECRET WAR BETWEEN DOWNLOADING AND UPLOADING

TALES OF THE COMPUTER AS CULTURE MACHINE

The MIT Press Cambridge, Massachusetts London, England

Designed by Brian Roettinger (Hand Held Heart), Los Angeles
The text was typeset using URW Grotesk, Sabon, and Zinnig.

Printed and bound in the United States of America.

Library of Congress Cataloging-in-Publication Data

Lunenfeld, Peter.
The secret war between downloading and uploading : tales of the computer as culture machine / Peter Lunenfeld.
p. cm.
Includes bibliographical references and index.
ISBN 978-0-262-01547-9 (hardcover : alk. paper) 1.Computers and civilization.
I. Title.
QA76.9.C66L859 2011
303.48'34—dc22 2010036048

10 9 8 7 6 5 4 3 2 1

For my daughters, Kyra and Maud

CONTENTS

ACKNOWLEDGMENTS

This book owes a debt to the theorists, artists, scientists, designers, programmers, architects and poets I was lucky enough to engage with over the years. Whether through *mediawork*: The Southern California New Media Working Group, the Mediawork publishing project, my academic seminars at Art Center and UCLA, or visiting lectures and critiques, these interactions helped me construct my models and challenged my preconceptions. I have had wonderful students through the years, and their receptivity to these ideas when they first were being formed greatly encouraged me.

A Faculty Enrichment Grant and sabbatical leave from Art Center College of Design were instrumental in moving this book from concept to manuscript. I took that sabbatical in Paris, and it was paradise to have an uninterrupted stretch of time writing at the Columbia University Institute for Scholars at Reid Hall. My thanks to Danielle Haase-Dubosc, Mihaela Bacou and Naby Avcioglu. Back in Los Angeles, Mimi Ito's Digital Culture Group meetings at the USC Annenberg Center for Communications were seminal for refining my arguments. Art Center's Media Design Program faculty, especially Anne Burdick, Andrew Davidson, Denise Gonzales Crisp, Brenda Laurel, Lisa Nugent and Philip Van Allen, were very supportive and their work continues to inspire me as to the computer's impact on practice. When I moved to the Design Media Arts department at the University of California, Los Angeles in 2008, I found congenial colleagues and a welcome new home.

My first year, I was invited to give three linked talks on campus to introduce my work to the UCLA academic community. Thanks to Leah Lieverouw of the Information Studies Colloquium, Kenneth Reinhard from the Program in Experimental Critical Theory, and Todd Presner and Jeffrey Schnapp of the Mellon Seminar on the Digital Humanities.

Paul Mathias graciously translated and published a long version of the talk I gave to his seminar in Paris at the College International de Philosophy in their journal *Rue Descartes*. Other venues publishing early versions of these ideas included *Afterimage*, *New Media & Society*, *Los Angeles Times* Sunday Book Review, *Think Tank*, and *Re:Public*.

Brian Roettinger of Hand Held Heart has a remarkable vision of how the book as object and system fits into the contemporary mediascape. Working with him was a reminder that the best design is not problem solving, it is situation producing. Chandler McWilliams contributed his expertise and vision to the electronic versions of the book. At the MIT Press, I continue to value the contributions of Deborah M. Cantor-Adams. A special word of thanks for Doug Sery, my long time editor, who has been so instrumental in building the field of new media studies. This is the ninth book we've worked on over the past decade, and as ever, he goes far beyond the call of duty.

Friendship sustains you in long projects, and so I want to thank Doug Hepworth, Dan Harries, Jeff Strauss, Ken Goldberg, Steve Mamber, Lev Manovich, Geert Lovink and Norman Klein. In the course of writing this book, my children Kyra and Maud grew from being delightful kids to insightful teenagers. Watching them gives me surety that the future will be a better place. Finally, there's Susan Kandel, my wife, my toughest editor, my model for writing. There are not thanks enough in the world to express what I owe her.

INTRODUCTION
THREE SIBLINGS

▶ notes: pp. 180

Sibling Rivalry

I am an only child with three siblings: the bomb, television, and the computer. I was born in the early 1960s, and these three have been part of my family for as long as I can remember. All of them came into the world decades before I did, in the crucible of World War II, but they surged to prominence at different periods during the second half of the twentieth century. As we hurtle into the twenty-first, this book is a call for the third sibling, the computer, to save the family from itself.

The bomb put an end to World War II, but inaugurated both the cold war and the looming fear that the fate of the earth hung in the balance. The bomb has been cited as the catalyst for everything from the rise of existentialism in the 1950s to the rebirth of religiosity in the West, from atomic age googie diners in Southern California to the *kawai*/cuteness of contemporary Japanese Superflat art. I was a baby when President John F. Kennedy appeared on all three major television networks to announce that the Soviets were stationing ballistic missiles just ninety miles off the Florida coast, and that what was later to be called the Cuban Missile Crisis was at hand. Kennedy used the medium of television to talk about the bomb, not only to the American people, but also to the leadership in Havana and Moscow, bypassing the customary diplomatic notification procedures entirely.

The family lore is that my parents stayed up all that night in terror for themselves and for me. All throughout my college years, I would occasionally look over my shoulder to see if there was a vapor trail in the sky pointing the way to atomic apocalypse. The history of the bomb is imprinted in our deepest reptilian brain; it is a history of fear, mutually assured destruction, and a blinding light followed by darkness.

If the first sibling came out as the biggest bully that the world had ever seen, what of the next one to emerge on the world stage? I may not be the best person to answer this question. A neighbor of mine wrote one of the first antitelevision books, *The Plug-in Drug*, and for years my parents didn't even allow me to watch the tube. But I outsmarted them, making friends with kids whose parents were not nearly so averse, and so the situation comedy double take is as embedded in my DNA as it is in any other red-blooded American's—even more so perhaps, as I eventually, and for reasons still obscure even to myself, went off to get a PhD in film and television. When I was very young, television was still something that you could contain, something that you could cordon off as separate from the rest of culture. By the time I was an adolescent, however, television had become the dominant medium, the all-encompassing ether in which everything that seemed to matter—entertainment, sports, news, politics, and even war—was suspended. Television was not a part of culture by the time I was in my twenties; it was culture. Television was the entertainer, with its twitchy history imprinted on our ganglia.

What of the third sibling, that late bloomer, the computer? While the other two were riding high—the bomb terrifying, and the tube distracting—the geeky third sibling was biding its time in university labs and high-end office parks, waiting for its moment. The famous and probably apocryphal comment from the 1950s that the world would need no more than a dozen or so computers points to the slow adoption curve in the general population. By 1982, though, *Time* magazine took the odd step of naming the computer its "Man of the Year" (giving me at least one excuse for anthropomorphizing these machines). In the go-go 1990s, those desk-bound machines and lumbering laptops were being connected into the global network that now defines the computer as the dominant sibling. This book is an announcement that the third sibling has truly arrived, and that if we play our cards right, there is a chance that it will not just eclipse its siblings but instead transform the world that they made. This is because the computer is the first media machine that serves as the mode of production (you can make stuff), means of distribution (you can upload stuff to the network), site of reception (you can download stuff and interact with it), and locus of praise and critique (you can talk about the stuff you have downloaded or uploaded). The computer helps people to create experiences and offers them spaces—often virtual and sometimes augmented—to share them.

Think of those fleeting moments when you look out a plane's window and realize that regardless of the indignities of contemporary, commercial air travel, you are flying, higher than a bird, moving through the air itself at hundreds of miles an hour, an Icarus safe from the sun. Now think of your laptop, thinner than a manila envelope, or your cell phone, nestled in the palm of your hand, or better yet, your ear. As computers get smaller, more ubiquitous, embedded in ever-more quotidian objects, faster, better connected, and easier to use, take a moment or two to wonder at the marvel.

You are the lucky inheritor of a dream come true.

The second half of the twentieth century saw a collection of geniuses, warriors, pacifists, cranks, visionaries, entrepreneurs, great successes, and miserable failures labor to manufacture a dream machine that could function as a typewriter and printing press, studio and theater, paintbrush and gallery, piano and radio, the mail as well as the mail carrier. Not only did they develop just such a device but by the turn of the millennium they also managed to embed it in a worldwide system accessed by billions of people a day.

While other technological dreams that sprouted up in the twentieth century—that *Popular Mechanics* future of flying cars, robot butlers, and thousand-story skyscrapers—never made it, this vision of a machine that can simulate any other is now a widely shared reality. Teenagers watch videos on their cell phones, ubiquitous grids create wireless hotspots in the middle of medieval towns, and interactive installations can be found in galleries worldwide.

The computer is a dream device, the first media machine that serves as the mode of production, means of distribution, and site of reception. It is the twenty-first century's culture machine.

But for all the reasons that there are to celebrate the computer, we must also tread with caution. This is because we are engaged in a secret war between downloading and uploading, and its outcome will shape our collective future in ways we can only begin to imagine. The promise of the third sibling, the computer in its guise as culture machine, is to build the feedback loops that keep this a virtuous rather than vicious cycle, but to do so will entail our taking sides in a secret war that is already decades old.

Before exploring this secret war, I would like to offer a word about the structure of this book. Interspersed within the body of the text are sidebars, offering stories of exemplary people, objects, and places. Here are synthetic musicals and silk books, cosmonauts and urbanists, brilliant samplers and the misguided nephews of genius. These smaller, more personal narratives will help to ground the theoretical and critical material elsewhere in this book. The question can reasonably be asked, Why does the critical and theoretical material require grounding in the personal? I would respond that the kind of universalized, omniscient theory we associate with the generation of 1968 does not connect with a culture that has been weaned on Oprah Winfrey's theater of confession. Winfrey's immensely popular television talk show ran for more than two decades, and within its syndicated, one-hour, afternoon format, Winfrey championed a relentlessly personal narration via mass media. By that I mean the ascent of the individual as creator and promoter of his or her own story, which is often but not always a narrative of redemption through suffering. The Oprahization of discourse has seen the decline of disembodied experience and ascendance of the memoir. Adjusting to the expectations of audience for the personal and the detailed is not, in the end, such a bad thing (perhaps the personal is political after all, and maybe even theoretical).

In addition to the main arguments and these sidebars, there is a third component to this book: a historical narrative that offers a generational history tackling the story of how the computer became our culture machine. The "Generations" section can be read before, during, or after the rest of the book. Like the sidebars, the historical narrative concentrates on personal stories, with two figures from each generation discussed at length, and a concentration on the ways in which the memes of simulation and participation developed and intertwined over the years. The first generation, the Patriarchs, established these foundational memes in the early years after World War II. They were followed by the Plutocrats, who turned computing into a business during the 1950s and 1960s. In opposition to the profit-minded Plutocrats, the 1960s and 1970s brought us the Aquarians, who proposed the visual, personalized, networked computers. In the 1980s and 1990s, the Hustlers took this vision and turned it into a commodity, getting it on to desktops worldwide. The next generation, that of the Hosts, connected these machines together into a truly World Wide Web, and pushed participation to the next level. We are now living through the sixth generation, that of the Searchers. For this generation, the wealth of information produced by the braiding of simulation and participation is so great that merely finding our way through the morass has become a signature quality of our engagement with the culture machine.

While the sidebars and "Generations" concentrate on people and discrete objects and systems, the body of the book engages with more overarching investigations that generate not just new findings but also new ways of talking about these findings. The pages that follow play with language, with pairings meant to establish complementary and oppositional relationships, like the central "downloading/uploading" coupling, along with "meaningful/mindful" as well as "tweak/toggle," "power/play," "simulation/participation," and "figure/ground." In these pages, there are also concepts that I proposed over a decade ago and am pondering still, such as "unfinish," "hypercontexts," and the computer as our "culture machine." Even newer words and phrases had to be coined just for the arguments presented here, like "Web n.0," "R-PR" (really public relations), "MaSAI" (or Massively Public Applications of the Imagination), "bespoke futures," "89/11," and "info-triage." Also present is the appropriation of language that happens when you scour science for concepts like "strange attractors," or when you create portmanteaus such as "plutopian meliorism" and posit that we can now speak of the "Enlightenment Electrified." Then there is the final issue of what kind of language differentiation you need to use in the face of a hybridizing hegemony of "unimodern unimedia."

Of special note in this book is the period between 1989 and 2001, in which all three siblings reached something of a tipping point. After the Berlin Wall came down and the sense of nuclear menace diminished, I stopped looking over my shoulder for the first time, expecting clear skies without vapor trails. But the events of 9/11 transformed the H-bomb into the human bomb, and the specific threat of death from the sky transformed itself into a free-floating anxiety about weapons of mass destruction and terror. At that point, television refined new ways of marketing fear as entertainment in a twenty-four-hour news cycle, and the worst excesses of the blogosphere simulated this model, accelerating it into the viral torrent of RSS feeds to mobile phones and "the new" at the click of the browser's refresh button. When fear or its inverse—empty-headed distraction—become the default content, the secret war between downloading and uploading is well on the way to being lost.

This book offers a warning, because if the bomb is the bully and the television is the entertainer, the computer is the family's mimic. And if the computer chooses to model its behavior after its siblings, we will be in worse shape than ever. This mimicry, or simulation as it is better termed, comes to us from the computer's very origins and will define its future. But should we push the computer to its limitless limits, taking advantage of its capacity to enable participation, we may well be able to address some of the key problems we face and make the first half of the twenty-first century more livable than the second half of the last one.

CHAPTER ONE

SECRET WAR

▶ notes: pp. 180–182

Humans Upload

First, we must define the terms of the struggle.

Downloading means pulling data into a system, and connotes moving information from a main or central source to a peripheral device. Uploading, by contrast, carries associations of moving data not only from a periphery to a core but also from one device to many, flattening out the hierarchy of production, distribution, and reception.[1]

All animals download, but only a few upload anything besides shit and their own bodies. Beavers build dams, birds make nests, and termites create mounds, yet for the most part, the animal kingdom moves through the world downloading and then munching it bits at a time.

Humans are unique in their capacity to not only make tools but then turn around and use them to create superfluous material goods—painting, sculpture, and architecture—and superfluous experiences—music, literature, religion, and philosophy. Of course, it is precisely the superfluous that then comes to define human culture and ultimately what it is to be human. Understanding and consuming culture requires great skills (ask anyone who has taught a child to read), but failing to move beyond downloading is to strip oneself of a defining constituent of humanity.

For all the wonders of the present moment, a cultural hierarchy persists. Even after the advent of widespread social media sites, a pyramid of production remains, with a small number of the members of a Web community uploading material, a slightly larger group commenting on or modifying that content, and a huge percentage remaining content to download without uploading.[2] One reason for the persistence of this pyramid of production is that like countries or peoples, different media have their own unique cultures. I would maintain that for the past half century, first the United States' and then much of the West's culture has been defined by television, and television is defined by downloading.

Television as a media system involves taking in images and sounds produced by others. It does not matter if it is delivered over the air, via cable, or with the aid of a dish; played back from tape, digital video disc (DVD), or a digital video recorder's (DVR) hard drive; watched on a plasma screen, an ancient console, or in the car (a particularly terrifying development for those of us who drive the freeways). Television is always the same: to watch it is to track an electronic download in real time—a narrativized progress bar with a laugh track.[3] Marshall McLuhan was half right: the medium is the message, but the messages also define the medium.

And what of the computer? The challenge it has mounted to television over the past decade has little to do with one machine being replaced by another—in the manner of 78s being supplanted by LPs, vinyl records by 8-tracks and cassette tapes, and compact discs (CDs) by MP3s; or videotape recorders by laser discs to be followed in turn by DVDs, video on demand, and DVRs. The challenge is far more profound than that. The computer, remember, is a machine that can upload anything its users make, and then distribute them either one to one or one to many, affording a radical break from the culture of television. But the computer also has the unique capacity for simulation, and it is this capacity—however perversely—that imperils its potential, because it can be better and faster at downloading than television ever was.

Let us turn for a moment to the cultural inheritance of television's half century of dominance.

Cultural Diabetes

The destiny of nations depends on the manner in which they are fed.
—Jean Anthèlme Brillat-Savarin

For diabetes sufferers, the body cannot create enough insulin to process the sugar that it has taken in: there is an imbalance between consumption and production. Diabetes is to a large extent a disease of plentitude, the result of obesity and the overconsumption of calories.[4] It is hardly news that we have been fattened up by a food industry that values novelty over nutrition and profits over public health. But in terms of our media consumption, we are in a diabetic coma as well.

The kind of diabetes I am refering to here is not type 1 (insulin-dependent or juvenile-onset) but rather type 2 (adult-onset) diabetes, which in past years has been hitting people at younger and younger ages. This type of diabetes is, in fact, largely treatable without drugs. A large portion of the affected population can keep their blood sugars in a healthy range without oral medications or injecting insulin if they follow an exercise plan, and eat in a way that controls the size of their portions and spreads out the amount of carbohydrates consumed throughout the course of the day.[5] In other words, the cure is under the individual's control, but individuals have to take responsibility for their own care by adopting a new rigor.

There are many conflicting theories about the massive increase in the weight, waistlines, caloric intakes, and incidence of diabetes in all age groups in the developed West. One thing all agree on is the confluence of syndromes, many of which are out of the individual's capacity to control. These include the shift from manual labor to desk jobs, from pedestrian cities to automobile suburbs, from home-cooked meals to restaurant or "take-home" consumption, and the tendency to "supersize" portions as a way to entice consumers into purchasing. But I have been attracted to one hypothesis that points to the adoption of high fructose corn syrup (HFCS) as the food industry's sweetener of choice over the past few decades.

For untold centuries, if humans ate something cooked that was sweet, it tended to be made with either cane or beet sugars. Starting in the 1970s, though, agribusiness invested heavily in shifting to a corn-based sweetener solution.[6] This worked spectacularly well for the food industry because corn syrup was much cheaper to produce, which allowed for heavily sweetened products to be maintained at very low price points. By the mid-1980s, almost all soft drinks, mass-produced bakery products, candies, and the like had shifted to HFCS. The move to corn sweeteners made the combination of a foot-long microwave cinnamon churro and a forty-eight-ounce Mountain Dew Slurpee® not just affordable but instead downright cheap. Supersizing has become not just one potentially viable economic model; it is increasingly *the only* economic model. The increasing availability of inexpensive high-calorie foods means that hunger is disappearing in low-income communities. Yet at precisely the same moment, diabetes-related problems are growing. So too the proliferation of ever-more opportunities to download is a gift that must be treated with care.

The ability to connect to networks at all times from anywhere can be a boon as well as an increasingly global promise that people can access the information they need. But the development of capitalism over the past half century was predicated on shifting patterns of consumption to concentrate on *wants* rather than *needs*. As a result, downloading has become yet more firmly intertwined with consumption.

These transformations are cultural, to be sure, yet they also rest on technological bases. Most commercial networks have radically slower speeds to upload material than to download it. So pervasive are these differences that numerous Web sites have come into existence that provide tools to measure them—important metrics for those who upload for a living to evaluate service providers. These disparities are not in and of themselves an insurmountable problem, but the infrastructure does build in a bias against the culture machine's capacity to upload.

This bias brings us back to the metaphor of cultural diabetes. Created like colas and burgers by multinational conglomerates, the junk culture of broadcasting creates a nation of intellectual diabetics. The cure is in our collective grasp. It involves controlling and rationing our intake (downloading), and increasing our levels of activity (uploading.) Not to break it down too much like a junior-year hygiene class, but what I am saying here is that watching is ingesting is downloading and that making is exercising is uploading. This project, then, sets as its purpose the identification of a new culture machine for the twenty-first century—one that uses digital technologies to shift us from a consumption to a production model.

▾ **SIDEBAR**

Ragazzi at Pizza Hut

Why are Roman teenagers eating at McDonald's and Pizza Hut, and what can patriotic Italians do to stop them? If marketing campaigns and the lure of the exotic convinced these *ragazzi* to scarf down frozen, prefab, meat-and-cheese-product-bedecked circles and triangles of carbohydrates, what hope might there be for the future of one of the world's most spectacular culinary heritages? When Carlo Petrini watched a McDonald's open in Rome in 1986, he understood that for Roman youth, U.S. fast food was both a symbol of modernity and an emblem of solidarity with teenagers the world over. Rather than merely wringing his hands, however, Petrini decided to take action. Bringing together foodies with anti-globalists, cultural traditionalists, and those who wanted to buy from local producers rather than multinational conglomerates, Petrini started what has come to be known as the Slow Food movement.

The central concept of Slow Food is that eating is part of an interlinked cultural system of production and consumption—gastronomy, in a word. The Slow Food movement is a rebuke to mechanistic visions of food as a commodity. The movement insists that when food is reduced to either fuel or instant gratification, people lose touch with the ways in which the practice of cooking and eating can become a way of life, a mode of culture. In this reconnection with the importance of daily practice, the slow food movement serves as a model for anyone who wants to think of moving out of a cycle of consumption for consumption's sake to one in which consumption is enmeshed in practices of production as well.

In the slow food movement, which has spread worldwide and now claims three-quarters of a million adherents, the rigors of learning to cook along with carving out the time for it are rewarded by the social interaction with the family, friends, and neighbors with whom one shares the experience. Beyond my own preference for gnocchi over nuggets, the slow food movement can serve as a model of resistance to television's junk culture more than the usual call for alternative, independent, community, and activist media (though it is interesting to note that Petrini himself comes out of leftist political media, having founded the first independent radio station in Italy). Its appeal may be the immediacy of food's place and moment of consumption in relation to production, and the surety that people have that the food they make will have an audience (all but the worst cooks will be able to find someone to eat what they have made). Yet the success of the Slow Food movement has had influence in other arenas, with calls for Slow Design, and Slow Architecture, and finally, the less organized but no less necessary Slow Sex movement.[7]

Television = The HFCS of the Mind

To claim that fifty years of television's dominance has given birth to a contemporary junk culture is to oppose the seeming consensus that television is in a new golden age of complex dramas, sophisticated niche comedy, and comprehensive documentary work. Any medium that has undergone the kind of explosive growth that television has over fifty years is bound to produce some interesting work, but to accede to its presence in our lives as an unassailable good is either naive or calculating, as people are always happy to hear that what they are already choosing to do is the best possible strategy. I call this strategy "capitulationism."

Often invoking "quality shows" like *The Sopranos* and *The Wire*, the capitulationists wax on about narrative complexity, visual sophistication, time shifting via DVRs, the release of whole seasons on DVD, and the increasingly intertwined hypercontextualization of television via extratextual material on the Web, including podcasting and mobisodes on mobile phones, all to make the claim that television has finally reached a critical mass of cultural importance. Yet if the formulation that the medium is the message holds true, the unfortunate fact is that the medium has not turned out to be all that good for us in heavy usage, even if some of the programming is as good as contemporary film.[8]

Television's junk culture spews the high fructose corn syrup of the imagination, and as a result of our addiction to the box, we have contracted cultural diabetes. I am fully aware of the critiques of my antitelevision position from both the Left and the Right, but before we get to the cultural studies rhetoric of audience empowerment and the laissez-faire bromides about the market, let me unpack my metaphors, first with some diagnostics, and then with some examples.

To begin with, for most people there is no more cinema. Cinema is simply the large format in which DVDs come out first.[9] In fact, cinema is no longer a mass medium; it is a niche medium with an audience substantially better educated and richer than television's—an elite medium, as absurd as that sounds looking at the offerings at the local multiplex. From its inception, the cinema was an urban, agglomerative medium that brought people together as an audience, and for decades its narratives taught us moral lessons about cohering as a population. The cinema might have looked back in Westerns and the occasional pastoral, but through the end of the 1940s, one of its central motifs was that of living together in the city.

Television, on the other hand, is a medium that atomizes general audiences. It breaks them off into either family units or, increasingly, individuals. Television regularly reflects on and leads its audiences into the suburban fold. The 1950s' trickle of television built on radio's earlier, opening of the home to broadcast media. By the start of the twenty-first century, what began as a trickle has become a torrent.[10] Television is a one-way spigot of privatized media gushing 24/7 into the home, commercial spaces like restaurants and supermarkets, and even schools.[11] The hardest task that television asks of its viewers is turning the power off after they have turned it on. This reminds us of what was was obscured in the bubble and bust of the last few years: the development of networked computers offers the first chance in a half century to reverse the flow, to upload and direct cast, rather than to download.

Patio Potatoes

The televisual era's twenty-four-hour, multichannel flow of entertainment into the home moves in only one direction, and the constant consumption of media without a corresponding productive capacity has engendered a sick culture. I might have used the metaphor of the perfect storm, but people know when they are getting rained on. In this case, we are more like the mythical frogs lounging in a pot of cool water who do not notice as the burner is turned on and they begin to boil.[12] We have been ignoring the heat, and now both our bodies and minds are at risk.

Philip K. Dick was typically prescient about the dangers of a culture of downloaded entertainment. He wrote that the "bombardment of pseudo-realities begins to produce inauthentic humans very quickly, spurious humans—as fake as the data pressing at them from all sides."[13] When AOL launched an Internet television initiative, its marketing executives were charged precisely with the bombarding of humans with pseudoreality. Describing how they would turn the wireless broadband computer into a mobile television, these executives discussed the importance of habits of mind: "We're looking to build behavior of viewing video online."[14] They rolled out their icon for this new service on the Reuters Spectracolor Board in Times Square and a supergraphic building/billboard on the Sunset Strip just before you drive west into Beverly Hills (where all the television people live). The image was of an anthropomorphized spud holding a laptop on a chaise lounge, and the tagline read, "Be a Patio Potato." We now know what to feed the Electric Sheep that Dick's androids dreamed of: patio potatoes.[15] Yet there is no reason

that the patio potatoes must prevail. Let us not forget that it took centuries to move from the local and artisanal production of food to large-scale agribusiness. The very flexibility and speed of change built into digital networks offers a positive note: it could take a much shorter period of time to head off the download-heavy moment. After half a century of television, where the habits of cultural consumption dominated, we now find ourselves supplied with a vast new infrastructure for uploading. We simply need to find the will to make the best use of it.

For me, the tipping point came in the midst of some random Web searching on one of the great thinkers about Southern California, Reyner Banham, author of *Los Angeles: The Architecture of Four Ecologies*.[16] Banham, an Englishman who admired freeways, claimed he "learned to drive in order to read Los Angeles in the original," made a documentary for the BBC in 1973 called, naturally enough, *Reyner Banham Loves Los Angeles*. I had never seen this video, and decided to search one day to locate an archive that had it near me. Then the whole video popped up, ready to watch in all its obscure, idiosyncratic, architectural, theoretical splendor. The Web blends the library, cinematheque, broadcast archive, and public square, and rendered the lot of them ever broader and deeper.

▾ SIDEBAR

A Brilliant Fiasco

Not all analyses are built from positivist research. Some evolve out of lyrical epiphanies. So prior or even a priori to a discussion of that most contemporary of machines, the computer, I would have us consider the *Livre de prières tissé d'après les enlumineurs des manuscrits du XIVe au XVIe siècle.* For the volume is not only an exquisite nineteenth-century reinterpretation of the medieval book of hours, it is also the unknown—and unknowing—origin point for contemporary screen culture.

Manufactured in Lyon by A. Roux between 1886 and 1887, this *Livre de priėrs* was the first and apparently only woven rather than printed book in bibliographic history.[17] Manufactured on the programmable Jacquard loom that enabled French industry to dominate the market for complex textiles, the *Livre de prières* was so intricate that it required hundreds of thousands of punch cards to produce.[18] It took A. Roux fifty tries to create the first salable version of this marvel of mixed technological metaphors, wherein Ariadne meets Gutenberg. The product of an industrial era, it simulated medieval content and anticipated by a century the information age. The book is woven of silver and black silk, and has a high thread count, allowing for crisp lettering and legibility. The effect of this tight weave, with its intense black figures on a silver ground, is remarkably like looking at a high-resolution display screen. The interwoven threads create the shimmer and pixilation of the cathode-ray tube screen *avant la technologie.*

Produced for the collectors' market, the book was considered "a brilliant fiasco," unable to sell out its first printing (perhaps better referred to as a first weaving) of sixty copies. Neither A. Roux nor anyone since has ever attempted to market another woven book. I shed light on this outlier to the history of information delivery devices as a way to demonstrate that even the most beautiful and seductive of technological artifacts may have no impact whatsoever on the culture at large, and that when technologies do effect a vast impact, it is because of far more than technological innovation or marketing savvy.

CHAPTER TWO

STICKY

▶ notes: pp. 182–185

Best Use

No one uploads more than a tiny percentage of the culture they consume. This is in contrast to conversation, which assumes give-and-take, and even religion (think of personal prayer versus the time spent in sermons). Of course people will download. Writers like to read, musicians listen obsessively, and game developers are above all players. But the goal must be to establish a balance between consumption and production, and using the networked computer as a patio-potato enabler, download-only device, or even download-mainly device is a wasted opportunity of historic proportions.

Shifting from consumption to another model is, rather obviously, to challenge the whole of our cultural infrastructure, not to mention an economy based on wants rather than needs. The great recession that began in 2008 provoked questions about the consumer debt-driven economy of hyperconsumption, but the moment that the economic indicators went back up, these kinds of discussions were completely abandoned. To question consumption has returned again to the fringes of discourse.

In any case, what makes me so certain that the culture warrants this attack? The market offers a solid rejoinder. Capitalist economies produce a range of media, and with their purchasing power, audiences choose what they want to support. To attack the state of culture is to attack the people themselves. This argument creates the classic horseshoe effect, bringing people from the libertarian far Right together with those on the Left who champion vox populi no matter what it is saying. The position of the latter goes something like this: no matter how debased the content, if viewers through the alchemy of fan culture repurpose it into a new mass/pop culture, it is a positive.

Even more perverse are those who claim that television is in fact a pedagogical boon—preparing us for an ever-faster twitch culture to come.[1] Those making these arguments claim the mantle of McLuhan without accessing the transgressiveness that made him exciting almost a half century ago. All assertions to the contrary, adopting the position that what people already consume is good for them has a huge audience in a world waiting and willing to embrace enabling wordsmiths. There is no need to capitulate, however, as the Web offers, as we have already seen, a whole range of deep content. It may not always be at the top of the list, or dominate whatever search metrics or ratings apply, but it remains the job of the critic along with committed audiences and makers to search for as well as support mindful downloading and meaningful uploading. Rather than capitulating, we are better off collecting "best-use" strategies for the use of these new technologies and aesthetics.

A century and a half ago, the English Victorian poet and critic Matthew Arnold demanded that critical inquiry ought to be "a disinterested endeavor to learn and propagate the best that is known and thought in the world, and thus to establish a current of fresh and true ideals."[2] Ever since Arnold's pronouncement, there have been furious battles about whether his lack of a concrete definition of what constituted "the best" doomed this declaration to be simply a reflection of his social prejudice and class position. As to his hope that poetry could save us and make us into better people, that too is either subject to debate or so far from the contemporary consensus that it is no longer worth the effort to argue. Rather than wade into any of these discussions on the moral character of culture's effect on the soul, I prefer to aspire to the best of Arnold's intentions, accepting that in our moment, categorizing "the best" is as much curatorial interventionism as it is a skirmish in the secret war between downloading and uploading.

Simulation and Participation

Any search for the best use of the culture machine has to engage with two, braided phenomena: simulation and participation. In computer science, the verb "simulate" or the noun "simulation" quite simply refers to the capacity to reproduce the actions, functions, and often "look and feel" of other computers, softwares, systems, and devices.[3] Simulation was important in the history of computer science because not all softwares were available on all platforms (they still are not, in fact), and being able to simulate meant that a single machine could behave like a set of others, all with different capacities and softwares. Since then, computers have been simulating not just other computers but also a huge range of other media. In the process of simulating photographic cameras—and their associated tools like developers and printers—the computer literally killed off its film-based model: in 2009, Kodak discontinued the iconic color film Kodachrome after three-quarters of a century, and once-dominant manufacturers like Canon no longer even manufacture thirty-five-millimeter cameras that take film. Video games may have begun in arcades, but they are now exponentially more likely to be played in the home than outside it. As for the cinema, which was itself swallowed up by televisual prostheses like videocassette recorders (VCRs), DVRs, and DVDs, the computer simulates it, migrates it online, chops it into YouTube segments, has it pirated on peer-to-peer networks, and shoots, stores, and projects it digitally. When computers simulate telephones, everything becomes available from the free Internet calling on services like Skype to mobile tele/computing hybrids like the iPhone.

When we are talking about communication devices, simulation engenders participation. After establishing communication between machines, between machines and people, and between people themselves, the next step is to allow the user to make something and then put it out into the network, where others will be able (and more crucially willing) to download that which has been uploaded. In other words, participation demands "affordances" from the system to move users beyond passive reception.

We inherit the concept of the affordance from industrial and then interface design. The usability expert and cognitive psychologist Don Norman drew from psychologist James Jerome Gibson, who was influential in changing the way we consider visual perception. According to Gibson, perception of the environment inevitably leads to some course of action. Affordances, or clues in the environment that indicate possibilities for action, are perceived in a direct, immediate way with no sensory processing. Examples include: buttons for pushing, knobs for turning, handles for pulling, and levers for sliding. Norman's immensely popular book *The Design of Everyday Things* moved these ideas squarely into the mainstream of industrial and especially interface design. His examples of "plates for pushing" and "knobs for turning" describe the typical course of interaction between a human user and a computer, or any kind of machine. During interaction, a user performs both physical and cognitive actions, and requires affordances to help with each. Norman calls these, respectively, real and perceived affordances.[4]

With media systems like television or digital media, we can think of affordances as everything from the development of better ways to interact with software and hardware (the graphical user interface comes to mind), to transformations in the conceptualization of how the hardware, systems, or softwares will be used in the world. These can be ideologically driven, market driven, or research driven. Usually increases in participation are driven by combinations of two or three of these agendas, rather than having one single force behind them.

Participation is what gets the power of computing and networks into living rooms. Here, I am talking about something more than "convergence." Convergence is when your personal digital assistant merges with your phone and adds in your music library. This Swiss Army Knife theory of technological improvement can be exciting, but the sheer inventiveness of the computer era will slow down if convergence is the ultimate objective. In other words, an end goal of simulating what already exists and then getting it out to as many people as possible is more limiting than it may at first seem. The point of participation is to be an active member of a vibrant, creative whole, instead of simply packing more and more media experiences into one little box (or a set of them strewn about the den).

When simulation evades the trap of mimicking the worst traits of a medium, and makes the best characteristics and affordances of it available to ever-larger groups of people, then simulation and participation become linked in what economists and social scientists refer to as a virtuous cycle. Should this virtuous cycle produce mindful downloading and meaningful uploading, then the promise of the culture machine is fulfilled.

▾ SIDEBAR

From Turing to Culture Machine

Computer science's equivalent to the Nobel Prize is called the Turing Award–an indication of how central Alan Turing is to the dream of the culture machine. A towering figure in a generation of truly great mathematicians, Turing was an authentic Cambridge eccentric, a shy but committed freethinker. He was by nature a solitary person, but proved to be a great patriot when he helped England and its allies crack German codes during World War II. Turing was, in fact, a perfect example of how both sides in the conflict harnessed the greatest minds of their generation to do both basic and applied research for the war effort. For his brilliant code breaking, Turing won the Order of the British Empire in 1945.

Written just before the war, Turing's master's thesis, "On Computable Numbers," was his greatest contribution to computer science. In it, he proposed the questions that still remain central to the discipline decades later. Turing suggested that it should be possible to make a "Universal Machine," a computer that could simulate the performance of any other device. The fact that the analog machines of the late 1930s and early 1940s were far too slow to function as Universal Turing Machines did not affect his faith that such devices would come into existence. And with the stimulus of the war effort, they did. Within a decade, Turing was working on the Manchester Mark I computer–one of the first machines recognized as being a direct antecedent to the computers we use now. Turing proposed a universal machine that functioned as a stored program computer; in this setup, the programs, or software, could be swapped and modified, improved and abandoned, just as the hardware could and would be. But in combination, hardware and software have become ever-more adept at simulating other machines.

In Turing's work we see the origin of a dream: a quest for universality and creative potential, a founding paper on simulation. Yet Turing was also involved in spreading the use of the machine beyond the technical fraternity. He assisted Christopher Strachey in producing what was probably the first artwork made with a computer: the love letter generator of 1952.[5] Strachey, working from a thousand-line piece of software (the longest yet written for the Mark I), created a program that randomly produced such sentimental and vaguely meaningless missives as:

Darling Sweetheart,
You are my avid fellow feeling.
My affection curiously clings to your passionate wish.
My liking yearns for your heart.
You are my wistful sympathy: my tender liking.
Yours beautifully
M. U. C.

Here, the Universal Turing Machine simulates mawkish Victorian sentimentality by choosing from a database of prewritten phrases that it then arranges into syntactically correct but stilted English. This trifle, inspired at least in part by the renown of Christopher's uncle Lytton Strachey's 1918 portrait of a generation, *Eminent Victorians*, is the product of a stored program computer, and as such may well be the first aesthetic object produced by the ancestors of the culture machine. The love letter generator's intentional blurring of the boundary between human and nonhuman is directly related to one of the foundational memes of artificial intelligence: the still-provocative Turing Test. In "Computing Machinery and Intelligence," a seminal paper from 1950, Turing created a thought experiment. He posited a person holding a textual conversation on any topic with an unseen correspondent. If the person believes he or she is communicating with another person, but is in reality conversing with a machine, then that machine has passed the Turing Test. In other words, the test that Turing proposes that a computer must pass to be considered "intelligent" is to simulate the conversational skills of another person.

Turing was not able to pursue these ideas much further because the same government that was happy to tolerate his eccentricities and use his talents to decipher enemy communications prosecuted him after the war for his homosexuality—still a crime in England at the time—and put him on estrogen treatments, then thought to reduce the effects of the "perversion."

He died in 1954, his death ruled a suicide, but with a complication so heartbreaking that it bears repeating. Turing's favorite movie was Disney's *Snow White*, and he died from eating an apple poisoned with cyanide. He left no note, and there are those who believe he rigged a way of dying that would leave his mother, with whom he lived, with some suspicion that it was an accident (or even murder) rather than the suicide that it was ultimately ruled to have been.

Info-triage: Downloading Mindfully

In any conflict there are battle lines, and the war between uploading and downloading is no different. These lines, or vectors, are drawn between two sets of poles: mindlessness and mindfulness; and meaninglessness and meaningfulness.

Our daily lives and routines are so busy that focus is difficult to attain. That is why we have automatic responses and habits of attitude. But there are times when focus is called for and should be summoned; it is this attitude that we call mindful. Mindfulness is not so much an innate trait as a learned response to the world. Mindfulness requires rigor. It is a muscle that must be exercised lest it atrophy.

In downloading, however, it is mindlessness that dominates. This is the inheritance of television. As we zap from channel to channel, so we surf the Web. Caught in the technotrance, our malaise masquerades as activity. But the clock is ticking. We cannot idly jump from link to link forever, pursuing everything even vaguely of interest because, sadly, we do not have eternity. The infoverse may be infinite, but our allotment of

days is not. Acknowledging the disparity between that which demands our attention and the limited time window we have on this earth demands that we deploy mindfulness. Whether we look to psychology or Zen philosophy for inspiration, mindfulness insists that we actively choose as well as commit to the situations and experiences we download. Only this can save us from that sense of attenuated distraction that characterizes too much of our essentially passive interaction with downloading.[6]

What tools and strategies, though, will we adopt in attempting to pursue mindfulness?

The first is *info-triage*. Triage comes to us from the French verb *trier*, which means to sort or organize by quality. After a culling, the third tier of coffee beans, for instance,was known as *café-triage*. During the Napoleonic Wars, however, the term became associated primarily with medicine as the French battlefield surgeon Dominique Jean Larrey used triage to refer the evaluation and categorization of the wounded. World War I brought the phrase and strategy to U.S. troops.[7] Emergency medicine has used the term ever since. As much as we are conditioned to the use of the word by our exposure to Hollywood war films and television medical dramas, the term triage along with its attention to mechanisms of organized and thoughtful choice prompts me to return to its earlier incarnation. What we need are not only technological systems to perform info-triage but also new habits of mind and practices of daily life.

Info-triage is more art than science, as a practice that involves the weighing of options and measuring of time. We tend to think of time in relation to efficiency, yet info-triage is about more than job performance. It is not so much a quest for efficiency but rather a practice devoted to mindfulness, the culling of distraction in the search for meaning. Info-triage requires a certain vigilance and temporal awareness. Primarily, it is about weighing options in real time, understanding that our capacities for downloading are actually limited, and that the choice not to engage at all is as valid as a choice between options. Info-triage accepts the psychological insight that those confronted with a vast array of options are often less satisfied than those who select between a smaller set of alternatives. Option paralysis shades into paralysis by analysis, and both are exacerbated by the never-ending data flow. Technology comes to the rescue at certain points, as with aggregator softwares that help people more easily manage multiple profiles on proliferating social networks, but info-triage offers more: a fundamental metric to balance the opportunities afforded by the flow and costs of choosing one over another, and most important, encouraging the option of diversion.

Technologists seeking to manage their own info overloads came to refer to "life hacking," although their tactics and strategies tend more toward rendering the world they deal with more efficient, as opposed to challenging what they end up managing.[8] The online community Lifehacker.com seeks out and evaluates techniques and technologies for "streamlining" life, positioning itself as "self-help for power users."[9] Those posting on Lifehacker are aware of the inherent irony of seeking technological solutions to an overdose of technology, but they feel that some action is better than none. Challenging or perhaps extending this action is the "Freedom" software that was inspired by the trap of ubiquity: "When there's wireless everywhere, how do we really escape the Internet?"[10] For Fred Stutzman, a graduate student in information science, the idea was to create an application that the user could turn on to enforce turning off connectivity for up to eight hours at a time. The only way to disable the Freedom software is to reboot the computer—a step that the developer of this free software hopes will be onerous enough to dissuade all but the most desperate. The goal is to use that time freed up from connecting and downloading to write, program, and create.

With Freedom, info-triage functions as a work-around in the engineering or hacking sense. A work-around does not "solve" a problem so much as circumvent it, going around it in order to achieve a more optimal performance.[11] Spam, data storms, and the ever-growing technologies, systems, and content that flow past us will not be solved so much as managed, and info-triage is just the sort of work-around that the moment demands.

Disrupting Flow

To achieve mindfulness in downloading and meaningfulness in uploading requires disrupting the flow of media that surrounds us. As the media mutated, so did the way that its contents flowed to, through, and around us.[12] The DVR enabled people to time shift their programming far more easily than did the VCR. But as DVR users time shifted their way through commercials, businesses began to embed advertisements within the content. As consumers turned to video games or online entertainment, e-billboards were sold in sports games and pop-up ads moved into browsers. Now, as the cell phone makes people more mobile and ubiquitous computing fills the world with information spaces, commercial speech in the form of advertisements, signage, and subtle cues to consume are ever more stealthily embedded around us in the invisible infosphere through which we move. In other words, when broadcast channels lost their centrality, televisual culture seeped outside the box and infiltrated other environments.

Short of complete renunciation, it is impossible not to "go with the flow," at least some of the time, losing ourselves in it as if we were leaves in a stream. But there are ways to step outside the plentitude and, at least occasionally, carve out periods of mindful engagement. This is vital because while the flow may be limitless, our time and attention is not. Until and unless the prophets of posthumanism can make good on their promises of eternal life, we will be bound by our limits as well as our aspirations. This is to stress the importance of uploading as a habit rather than as a mere technological affordance.

Creating cultural hierarchies can make citizens of a democracy nervous. Who is the critic to judge the meaningfulness of experiences to other people's lives? Leave that sort of assessment to the individual or the market. Yet the networked computer has ushered in an era of exponentially increasing cultural production, which democratizes the ability to create at the same time that it impels us to create new ways of hierarchizing what we encounter.

When a new medium explodes on the scene, we have to find ways of responding to the demands on our time and attention. Strategies like renouncing an individual medium such as comic books, the Web, or even television as inherently evil are retrograde (though each medium has had critics completely condemning them).[13] Such a strategy ignores the richness and pleasure of contemporary work in favor of a fusty antiquarianism. The opposite strategy, a capitulation to whatever the market and the network throw at us, is nothing less than a mindless immersion. The technocratic search for an "efficient" use of contemporary media by using configurable and interactive tools to restrict what you see to what you define as your "needs," can result in tailored news reports as well as a numbing reinforcement of sameness. The possibility for serendipitous encounters with the new and challenging cannot be abandoned in the quest for time management. In the end, the issue is less of criticism than it is of curation—the marshaling of culture, the mindful juxtaposition of ideas, images, sounds, and interactions to create more than the sum of their parts.

▾ SIDEBAR

A Brilliant Success

We discussed the brilliant fiasco of the Livre de prières, but I would like to shift to the brilliant success of "Kitch's Bebop Calypso."[14] If the woven book was a starting point for contemporary screen culture, this obscure calypso recording from 1951 can be seen as a model for twenty-first-century creativity. A product of reverse continental drift, "Kitch's Bebop Calypso" collapses the Atlantic divide, mashing together Trinidad, Curaçao, Aruba, Jamaica, the British Isles, and faraway Manhattan to form a sonic Gondwanaland. Kitch was born Alwyn Roberts in Trinidad, and renamed Lord Kitchener in Jamaica after he came to dominate the local music scene. Calypso was the music of carnival and boasting, a way to celebrate cricket victories and bemoan infidelity. It was the public language of the first wave of black immigrants to London and Manchester after World War II. Lord Kitchener was on one of the first boats, and he can be counted as the first important musician of color to contribute to England's musical heritage.

Bebop was the new jazz that had come to supplant swing, which had been the sound track to the Great War. Bebop was Charlie "Bird" Parker and his saxophone, Miles Davis sporting sunglasses at night, and Dizzy Gillespie bending his trumpet into a new shape for a new sound. Bebop, the nickname of the flatted fifth that swing bands virtually banned, was the confrontational sound track of 52nd Street, a sound from Manhattan after midnight, arriving just at the moment that New York City became the de facto capital of the world after the long and terrible war.

What you hear in "Kitch's Bebop Calypso" is the excitement of discovery transformed into a blaze of creativity. Lord Kitchener was building the calypso scene in London—performance by performance, club by club, record by record—but he was also listening to what was going on in other genres and around the world. The calypsonian, known for his lyricism and smooth delivery, has his band start up with a lively horn arrangement, and he then begins singing:

Well I nearly went crazy / when I heard the record of Gillespie
It really enchanted me / just to hear him play "Anthropology"

The West Indian inflection that manages to rhyme "crazy" and "Gillespie" then gives way to what any twenty-first-century hip-hop fan would recognize as the drop of a sample: Kitchener literally plays a short selection from Gillespie's far more frantic "Anthropology" to make his point. He shares his enthusiasm for the music and even gives pointers to others for how to appreciate bebop—"If you listen carefully / you will surely / enjoy the melody"—a direct response to the old guard that claimed bebop lacked an identifiable beat, that it had abandoned the syncopation of swing to indulge soloists.

But Kitchener uses his own command over the calypsonian forms to assess these New Yorkers:

I'll give you the ratings / of different composers
Two recognized artists / Charlie Parker and the Miles Davis
Mr. Parker also plays / he's rating highly in the steeplechase
and Miles Davis again improved / when he made this number he called it "Move"

And here again, Kitchener drops the sample, creating a juxtaposition between the calypso beat and Davis's horn, to the advantage of both.

The listener gets the feeling that Kitchener was waiting at the docks for the boats to arrive and unload their new records from the United States, ready to open himself up to the new thing and share it with his own audience. Right there on the defunct Melodium label, Lord Kitchener was showing the way to use as well as recombine elements and fragments of culture in new and inventive ways. That he had a whole band behind him is immaterial to the immaterial sampling that followed in his wake with the advent of digital technologies.

If you listen to "Kitch's Bebop Calypso," the absolutist arguments against remix culture dissolve. Here is an artist delving into the sonic archive, embedding the fragments in a different though complementary context, and weaving the whole together as a joyous celebration of creativity and music. I have no idea if Kitch paid much attention to computers, but anyone who listens to his music will understand that he cared deeply about culture. I first heard "Kitch's Bebob Calypso" on Internet radio—a clear case of the computer simulating another medium. As a participatory act, I sent an MP3 file of the song to a disc jockey with whom I was working on book about remix culture. This is a microhistory of the use of the culture machine: simulation and participation meld to create the way we live now.

Sticky vs. Teflon

Meaning is like pornography. We recognize it when we see/read/listen/interact with it. This evasion brings us to the next stage of value judgments. If we can distinguish between mindless and mindful downloading, then we need to talk about uploading in terms of its meaningfulness. A way to expand

on the earlier definition of meaningfulness is to employ yet another term: *stickiness*.

Stickiness refers to surfaces, assemblages, and experiences to which other things adhere.[15] A truly sticky experience should offer the possibility of accumulation. This can be a new way of thinking about the concept we used to call depth, but that now needs to expand out in various ways. Stickiness is about creating and uploading media that can overlap layers of meaning, such that downloading the material creates experiences deeper than distraction. A sticky object or system has affordances that allow other meaningful objects or systems to latch on to it, expand it, or burrow deep within it. Sacred texts like the Bible and the Koran, classics like the *Iliad* and the *Odyssey*, the Sanskrit epics *Mahabharata* and *Ramayana*, and the plays of William Shakespeare have a vast amount of stickiness due to their long duration along with the vast body of textual analyses that each has generated.

Stickiness is a quality that can accrue to almost any human activity. By sheer obsessiveness and duration, it is possible to add stickiness to even the most meaningless activities and trivial pursuits. By this I mean the ways in which the obsessive collection of 1930s' Tin Pan Alley songs creates a window into a world, or the minute observations of schedules and deviations that the wonderfully compulsive English train spotters produce.

Rather than sticky, I would say that there is something about fans' appropriation of mass culture that produces what I would characterize as *Teflon* objects.[16] When fans appropriate the materials and aesthetics of commercial culture, and then make new things out of them, they are indeed "producing," but I worry that there are huge disjunctions between the level of personal investment and the substance of the output. I am not convinced that fan-produced media is all that sticky, except perhaps to other committed fans.

Meaning is a loaded word. In consciousness, everything has a meaning or can be assigned one, including the statement that meaning has no meaning.[17] In this context, I am not interested in discussing what philosophers like Martin Heidegger have called the fundamental question of meaning—Why is there something instead of nothing? —but instead want to think about the meaningfulness of works of culture.[18] This impels us to create a hierarchy of meaning, judging some things to be more "meaningful" than others, and sometimes going so far as to label them "meaningless" (though we know full well the impossibility of any *thing* in culture lacking meaning).

Meaningful uploading is both a counterpoint to and an ally of mindful downloading. As noted, any definition of what is meaningful runs the risk of tumbling down a rabbit hole of philosophical debate. Yet I am willing to stake a claim on the idea of the cumulative as one place to start. Work uploaded into the world ought to have enough of an affordance to connect with other elements of the network to add to larger questions of meaning rather than simply shimmering there as nodes in the distraction machines. As has been discussed, the issue is to create sticky rather than Teflon media, uploading work that accretes into meaning as opposed to bouncing around, atomized and distracting at best. Just as no one will download mindfully at all times, it is an impossible request to ask people to only upload meaningfully. But setting the bar too high is preferable to not setting the bar at all.

Fifty years ago, the categorizing of meaning was considered to be one of—if not the—chief calling of the critic. The advent of critical theories like poststructuralism, deconstruction, and postmodernism put many of the classic categories in jeopardy: building a canon around the good and the beautiful was "problematized," high and low ceased to function as viable categories for culture, and progress and truth were discussed as creations of power struggles rather than immutables in the human condition. Theory with a capital T practiced a brilliant negative dialectics, but did not always replace the overthrown concepts with new, more congenial ones. In the end, this dismantling of preexisting cultural norms helped to create a void that allowed commercial culture to reign unchallenged.[19]

Although we are talking about billion or even trillion dollar markets when we invoke commercial culture, many of the new affordances being built into mass culture for fan "participation" are equivalent to cafés at the mall: pleasant enough respite, yet still in the mall. Much ado was made of the fan community's contributions to the pre- and postproduction phases of the B movie *Snakes on a Plane* (2006). As a genre film and star vehicle for veteran tough-guy actor Samuel L. Jackson, *Snakes on a Plane* was bound to generate dialogue on message boards, especially after the trailer for the movie became a viral video success on the Internet. The sheer volume of excitement about the film and the vociferousness of the fans in chat rooms, blogs, and Web comment rooms, however, caught the director and producers by surprise. These "creatives," as Hollywood calls them, responded by opening a dialogue with the fans, modifying the script, and shooting new scenes based on the latter's input. All this took place prior to the film's release and was heralded by the capituationists as a new era in fan culture. But then the film opened, and all of that fan activity, as pleasurable as it might have been, was seen to have contributed to a film just as bad or even worse than other genre work with no fan input. Instead of living up to the ideal of a pulp masterpiece, the "collaboration" between the professionals and the amateurs produced nothing more than a subpar, Teflon-coated pellet.

The phrase "participatory culture" has been added to the lexicon to describe this kind of fan-driven work, distinguishing it from popular or mass culture.[20] But what is the point of developing these machines, networks, and affordances for the delivery and publishing of media if we don't also develop some corresponding sophistication in their content as well as their use? There are limits to what mass culture can talk about, and the levels of subtlety, language, and thought and thoughtfulness. The question of technique is harder to pin down, as mass culture has such economic might that the newest and most powerful of tools and techniques are always open to it, but of course mass culture tends to turn these techniques into clichés within just a few business quarters.

It has come to the point that we cannot imagine anyone other than a gifted artist making use of morphing technologies without reducing the audience to tears of laughter or pain. The commercial appropriation of twentieth-century political avant-garde techniques serves as a warning that there is no inherently revolutionary quality to new technologies and their aesthetics. The 1920s' montages of Soviet filmmakers like Sergei Eisenstein and Dziga Vertov ended up the standard tool for portraying boy bands on MTV in the 1990s. Surrealist art shocked the bourgeoisie in the 1930s and sold everything from soap to tampons half a century later. Deconstructed graphic design sparked the legibility wars of the 1990s, and by 2000, jagged type sells even the most innocuous brand concepts like the "Got Milk?" campaign. Street artist/designer Shepard Fairey, fresh from creating the iconic Hope poster image for Barack Obama's presidential campaign, was hired to craft Saks Fifth Avenue's spring 2009 marketing push, and did so with Soviet-era propaganda graphics proclaiming, "Want It!"

Historical precedent is a strong indication that no matter how "revolutionary" the possibility of mass participatory fan culture might have seemed at one point, that potential is not sufficient to support my aspirations for the culture machine. Just as Kingsley Amis so ruthlessly summed up his worldview with *Lucky Jim*'s observation that there "was no end to the ways in which nice things are nicer than nasty ones," there is no end to the ways in which sticky culture is stickier than Teflon production, no matter how participatory.[21]

Power and Play

In the more than half century since the computer emerged from the conflicts of World War II, there has been a transformation of our relationship to data, and in turn, to the ways that we generate meaning and content. As already noted, computers are machines unique in their capacity to simulate other media. When they are connected into a network, simulation joins with participation to define the culture machine. These simulation networks not only increase users' ability to manipulate a range of symbols and situations but also vastly broaden the pool of people who do so on a daily basis. From the scientist shifting the unknowns in an equation, to the designer cycling through background colors in a digital collage, to the schoolchild shifting a paragraph around in a draft of a paper, each of us has entered an era defined by power, on the one hand, and play, on the other. We have the power to effortlessly shift variables, and this encourages us to play with our data sets.[22]

Let us consider more closely the designer who changes the background colors over and over again in hopes of pleasing a client. Each time she changes the color, she is engaged in *tweaking* the system, using its power to play with the variables in real time. The active tweaking of processes defines our engagement with information technologies in the realm of the computer. Tweaking is both a result of and contributing factor to the information expansion (or explosion), which has been discussed widely enough to justify the common phrase "age of information." I am referring to tweaking in search of additive levels of meaning as opposed to the more formal or simply obsessive-compulsive tweaking described above.[23] There is no denying the ways in which computers exist in a continuously shifting and fluid blend of text and context. The issue is how to use this fluidity to build meaning rather than increase distraction.

Toggling is a related phenomenon. It refers to the effortless shifting between views of a data space. The quintessential toggle is between a first-person point of view and a God's eye, overhead mapping. Toggling between the two points of view has been central to everything from three-dimensional architectural programs to first-person video games. In the first case, the designer shifts back and forth to visualize the space as it is being created. In the second instance, players toggle between their own vision of the action and an overhead, aboveground view that helps them map out strategies and movements through the game space. Toggling continues to utilize perspectival techniques dating back half a millennium, but the ease of moving between multiple views creates a thoroughly contemporary affordance for the user.

Toggling and tweaking are both examples of how the computer effects quantitative shifts with qualitative results. Playing with perspective and point of view is as old as peekaboo games, but digital systems do not allow, so much as demand, that makers and users hide, show, and switch between elements, views, and screens. What begins as novelty—the earliest computer games to offer the capacity to toggle between a player's point of view and a God's eye view had a distinct marketing advantage—becomes an expected affordance of the next generation of not only games but also a huge array of electronic interfaces. As for tweaking's impact, it is important to understand that iteration, defined as a series of repetitive actions with successive changes leading to new results, has long been a part of the creative process. The computer speeds up iteration so furiously that the creative process is itself remade into what we will next discuss as an ever-unfolding aesthetic of unfinish. Tweaking and toggling are built into any system with enough power to allow for play, and the trick is to balance the temptation to tweak and toggle endlessly with the opportunities that these affordances offer for creative work.

Unfinish: Continuous Partial Production

Twitter, the microblogging software that limits users to 140 characters or less, distributes its messages via a range of delivery mechanisms to desktop applications, mobile phones, and instant messaging systems. Twitter promotes itself as a way for "friends, family, and co-workers to communicate and stay connected through the exchange of quick, frequent answers to one simple question: What are you doing?"[24] Twitter is something of a limit case for participation, and inverts a classic analysis of Web 1.0 interactions. In response to the explosion of new media forms in the 1990s—from cellular communication to video games to email to the World Wide Web—social computing expert Linda Stone coined a phrase that became famous: "continuous partial attention." Continuous partial attention differs from multitasking in that the individual "wants to be a LIVE node in the network . . . [to] be busy, to be connected, to be alive, to be recognized, and to matter."[25] Recent developments in the Web coupled with the broadening of social networking encourage us to invert Stone's coinage and move from attention to production.

The growth of Web logs, or blogs, considered in tandem with Flickr, Digg, and other social softwares that enable posting and tagging accounts, creates an environment that I categorize as "continuous partial production." As 99 percent of everything ever made is either purely for personal consumption, largely forgettable, or just plain junk, continuous partial production is not a huge problem. What does become problematic is when the new affordances make the old content untenable in the emerging environments.

Acknowledging that there are losses that follow every gain in technological capacity is not the same as blindly following the reporting cycle. The crucial issue is that these twenty-first-century cultural machines lead to a previously unimaginable level of object differentiation and information richness. The networked culture machine's combination of embedded technology and just-in-time production make possible a novel, hybrid intellectuality. Text can be linked to graphics, photos, and moving images in fluid ways impossible a generation ago. The combinatory potentialities of alphanumeric texts, still and moving images, aural components from music to spoken word, and even contextual environmental embedding—all of these simulations of other media—offer a huge set of affordances for both the creation and reception of meaning. The sheer density of information and materiality of the contemporary moment is unrivaled in history.

The key to making meaning with the culture machine is to harness the two defining modes of networked computing—simulation and participation—in order to add stickiness to the culture. One way to increase stickiness is to use the culture machine to add a quality of *unfinish* to its production. What an author produces is open to revision, and those who used to be readers, listeners, or viewers can become users, through appropriations, remixes, and creative reuse. The idea that everything is essentially an iteration can be terrifying because it encourages an endless tweaking, rather than a commitment to the discrete project with a beginning and an ending. Software developers occasionally refer to this as being in a state of perpetual beta, meaning that the code will never be released in "final" form, and is subject to a continuous process of review and reform. But the new era of unfinish can be used to acknowledge that every cultural product eventually relates to and is transformed by its contact with users and other products. As we will see later, acknowledging unfinish is critical to the development of open-source and Creative Commons approaches to producing, licensing, and distributing media.

Unfinish challenges authorial intent, on one side, and immutable meaning, on the other. Digital unfinish builds on twentieth-century cultural explorations of these issues, but the computer and the network transform the baseline assumptions. Objects can be produced that are open to later modification, which is a key attribute of open-source creativity, and for any damage that can be done, the chance is there to undo it—the Control Z that we wish we had in the material word. There are strong connections between unfinish and chapter 4's discussion of Creative Commons and the open-source culture movement that networks have made possible. Thinking about unfinish leads to questions about openness as well. One metric for the success of a technology, especially a digital one, is to look at how open it is to unanticipated uses. How unfinished is it? One of the valid critiques of modernism was that in its utopian fervor, it regularly discounted users' contributions to the design schema.

There is a story about Walter Gropius being questioned by a reporter from the *Harvard Crimson* about the Cambridge dormitory that he had just designed. All the furniture was bolted down, and the student asked what would happen if someone wanted to rearrange it to suit their own taste. The great modernist replied that anyone who would want to move anything would have to be "neurotic."[26] Networked society has overreacted somewhat to this modernist arrogance, though, embracing marketing's obsession with surveying the public to see exactly what it "wants" and then supplying it, anticipating the user's every need. This cravenness is unnecessary, and forms an expectation of perfect responsiveness that fits too closely into established commercial relationships. Strategies open to productive unfinish instead anticipate users who acknowledge that they live in a Heisenbergian universe.

In quantum mechanics, physicist Werner Heisenberg postulated the uncertainty principle, claiming that the presence of the observer changes the conditions of that which is being observed, and that it is thus impossible to have an "objective" view of any phenomenon. Accepting unfinish similarly implies that the introduction of any new technology changes the user's environment in ways that may generate entirely new models of use. This is unfinish as a dynamic, autocatalytic system.

An economy of unfinish shifts us from a pure consumption-oriented model to one that mixes production and consumption. It becomes an economy in which we produce tools for the creation of new objects, experiences, and communication. It is this emphasis on production that distinguishes it from consumption for consumption's sake. This stress on production also shifts our idea of what audiences should be. We have come to depend on a dichotomy of audience and artist. With eight-plus hours a day of television viewing, the imbibing of professionally generated entertainment has reached unprecedented levels. An era of unfinish and a move toward production/consumption will steer us away from the notions of aesthetic form that we inherited from the traditional arts, and take us to other modes and models of engagement with experience.

CHAPTER THREE

UNIMODERNISM

▶ notes: pp. 186–188

Unimodern Unimedia

There are an exponentially growing number of people who can not but see the world as information itself. This is the key to understanding the aesthetic effects of the culture machine. This chapter offers a larger vision of how the computer becomes the central node of culture itself. Artist John Simon Jr. rips apart the guts of a PowerBook to create a display space for the evolving software simulation of *Complex City.* Designers IOD craft their Webstalker software to give visual form to the sprawl of the network, and Lisa Jevbratt maps out the Web as an interactive color field. Aaron Koblin makes a live-action video for Radiohead's song "House of Cards" without cameras or lights, using 3-D tracking technologies that create data streams that viewers/users can then remix with new angles and visuals to post to YouTube. The green-on-black datascapes in *The Matrix* films simultaneously virtualize and realize the Wachowski Brothers' pop mysticism. Even Frank Gehry's Guggenheim Bilbao and Disney Concert Hall, those most sensuous of twenty-first-century signifiers, can be seen as manifestations of the CATIA 3-D software used to design them. Ubiquitous computing and geographic information systems are virtual figuring machines, constantly popping out new data points from previously mute spaces and maps. How are we to describe these products of the culture machine?

The culture machine was originally tagged as the ultimate in "postmodernism." The collage aesthetic, decentering, and obfuscation of both author and authority that the networked computer offered seemed like a perfect fit with those who saw the end of the high modern moment as being superseded by the postmodern. But I would argue that rather than early, high, or post, we produce and consume a *unimodernism.* Our moment is *unimodern* in the sense that it makes modernism in all its variants *universal* via networks and broadcasts, *uniform*

in their effect, if not affect, and *unitary* in terms of their existing as strings of code. In the unimodern era—as bits, online and in databases—a photo is a painting is an opera is a pop single.

▾ SIDEBAR

After Allie Mae Burroughs

Arrange three portraits on whatever screen is handy. They are at once the same and not the same, but it is their sameness in difference that will prove crucial to understanding our moment. There is a "subject" of these images who looks out at you, instantly recognizable though still anonymous. Her brow is furrowed, her lips pursed, her raw skin crisscrossed with crow's feet, her eyes dark and tired. When you find out that she is only twenty-seven years old, you are reminded of the toll that poverty takes on the body. "She" is Allie Mae Burroughs, a sharecropper's wife from Depression-era Hale County, Alabama. And through various quirks of fate, art, and technology, Allie Mae Burroughs remains relevant well into the twenty-first century—a posthumous heroine of unimodern unimedia.

The first image on our screen is a photograph taken of Burroughs outside her clapboard house. In 1936, Walker Evans, one of the towering figures of modernist documentary photography, accompanied the writer James Agee on an assignment to document rural poverty in the Deep South. Five years later, the project became a book with the evocative—and biblical—title *Let Us Now Praise Famous Men.* The book opened with dozens of Evans's images, including its most iconic, the photograph of Allie Mae Burroughs, though here Agee gave her the first of her new identities: the pseudonym "Annie Mae Gudger." An idiosyncratic mix of reportage, bellelettrism, and what might be called American secular theology, *Let Us Now*

Praise Famous Men was a critical success, but a commercial failure, selling just over six hundred copies in 1941 and then promptly going out of print. Reissued twenty years later and still in print half a century on, *Let Us Now Praise Famous Men* became a sensation, inspiring generations of aspiring photographers and journalists to go out and capture the truth of their world. The book also helped cement Evans as one of photography's great patriarchs.

The second image is a photograph of Evans's photograph of Burroughs. On our screen it looks exactly the same as its predecessor, yet when this "new" image was first shown in 1981 at Metro Pictures gallery in New York City, it was called "Untitled (After Walker Evans)," and credited not to Evans but instead to a young artist named Sherrie Levine. Levine's "rephotographing" of Evans's iconic image has been interpreted as everything from a blatant act of plagiarism to a daring move of appropriation. I side with the latter view, taking "Untitled (After Walker Evans)" as a quintessential postmodern work, using media to challenge media, authorship to challenge authorship, and dare I say "genius" to question what we mean by the term in a culture that almost never bestows it on women. The Evans estate, however, saw Levine's intervention as theft—it sued, won, and prevented the public presentation of Levine's series for more than two decades.

Return one final time to the screen for the last of our images of Burroughs. This one is a digital scan of the Levine rephotography, put online by the artist Michael Mandiberg in 2001 as part of a project called AfterSherrieLevine.com. The scans are at 850 dots per inch, the same resolution as Levine's work, though much lower than what could be done with the Evans "originals." Each can be downloaded from Mandiberg's site with a "certificate of authenticity" proclaiming that they are "real" Mandibergs—an ironic riposte to those who took Levine's work merely as the production of "fake" Evanses. Even if AfterSherrieLevine.com is less compelling than the work of Levine herself, it does provide a succinct reminder of exactly how unimodern unimedia function. Because when the three images of Burroughs are lined up in whatever format, be it desktop, laptop, on a cell phone, or floating in the ether as a retinal projection, their differences are wholly subsumed by their new, technologically mediated equivalences. Indeed, questions of authenticity, primacy, and ownership dissolve as the image is simultaneously autonomous and mediated, unitary yet boundless. Walker Evans is dead. So is James Agee. But Allie Mae Burroughs lives on—in the collection of New York's Museum of Modern Art (which purchased both the Evans and Levine photographs), not to mention on an infinite and infinitely proliferating number of (unlicensed) screens, near and far from you.

Figure/ground

The unimodern culture machine produces vast databases of texts, images, sounds, and other media. Downloading mindfully from this enormity, much less uploading to it in any meaningful way, requires, as we have been discussing in earlier chapters, the development of new habits. One way to develop these habits of mind is to think in terms of figure and ground relationships. Gestalt psychology's figure/ground experiments were provocative: the drawing that can be either a vase or two faces in profile; the rabbit that's a duck that's a rabbit; the old woman who is a young woman who is an old woman. When you look at an image, the figure is what is supposed to have the definite shape, the prominent contour, and to use the peculiar, Germanic phrasing beloved of Gestaltists, a greater "thing

character" than the ground. In the best of the figure/ground illustrations, there is a moment when your perception "pops," and what had been the ground flips instantaneously into the figure: it becomes "thingy." This is a transformation simultaneously magical and quotidian, and after the pop, it seems impossible that you ever did not see the figure in precisely that way.

This kind of pop can signal a major transformation in culture: some "thing" emerges from the ground, while formerly prominent figures sink back into the amorphous periphery. The dynamic between figure and ground is akin to a paradigm shift, but it is less about the singular figure exploding the system through invention than the collective recognition of things that were already present although not central to the culture's perception of itself. The pop between figure and ground is as contemporary as Malcolm Gladwell's notion of the "tipping point" and as ancient as the story of the emperor's new clothes.[1] A crucial difference between the revolutionary model and the figure/ground flip is that in the former, there is no going back, while in the latter the figure and the ground can oscillate. The great designer Paul Rand once claimed that his process could be described as an endless series of "figure/ground problems. Everything is! . . . All art is relationships."[2] These relationships are even more important in an era when toggling between figure and ground can happen at the touch of a switch, the shift of an electron.

After the Pop

One of the loudest pops of the last hundred years occurred as a result of the publication of Sigmund Freud's *Interpretation of Dreams*. Within a generation, we were no longer unconscious of the unconscious: the ground of unacknowledged motivations became the central figure for a culture in thrall to psychoanalysis. It is not that artists before Freud were oblivious to inexplicable motivations and hidden meanings—think of Macbeth's dammed spot—but rather that after Freud, those factors that formed the ground for artists from Aeschylus to Shakespeare to Honoré de Balzac were transformed into the very figure of the work. Uncovering the unconscious in post-Freud art and culture was easier than fishing with dynamite. Just think of the way that Alfred Hitchcock treated poor Jimmy Stewart in film after film: crippled, castrated, and frustrated. If you go looking for the unconscious in *Rear Window*, *Rope*, and *Vertigo*, how could you not find it? Psychologists call this "selective perception." Once you become aware of the unconscious, how can you miss it (or its absence), wherever you look?

After the pop, the new relationship takes on the mantle of "common sense," of natural perception and transcendent truth. We could with equal ease question how middle-class strivers in the 1920s could have demanded so much Victorian bric-a-brac in their homes when the machine aesthetic was so clearly ascendant. But that ignores the reality of the situation that confronted the first generations of modernists. The general population knew that industry and the machine were the ground on which their culture was built; they just did not want to see it as central. They wanted art nouveau curlicues, symbolist syphilitic femmes, and dead-earnest arts and crafts wallpaper. It was left to those small communities of the like-minded and loudly debating—the cubists, futurists, purists, constructivists, and other avant-gardists—to flip a world in which the industrial machine was the ground from which the figures of culture were drawn. Think of the incendiary industrial chic of Marcel Duchamp's fountain, the antihumanism of Le Corbusier's machines for living, the scopic mechanization of Vertov's kino-eye, Filippo Tommaso Marinetti's wholesale embrace of speeding planes and motorcars, and Otto Neurath's universalized design isotypes. Over the course of more than a quarter century, these artists forced us to confront the machine as the salient element of twentieth-century visual culture, not simply the backdrop.

The visionaries of the early twentieth century transformed the look and feel of culture, not supplanting the oak and marble edifices of the past so much as adding the sheen of industrial materials like concrete, glass, and steel. By the 1920s and 1930s, the audience for modernism was equally an audience for the machine aesthetic: the hard, unembellished lines of El Lissitzky's graphic design; the clanks and atonality of Alban Berg's opera *Wozzeck*; the sensuous curves of Marcel Breuer's chromed steel tubing in his "Model B32" chair; the assemblages of tubes, pistons, and levers that compose the Fernand Léger painting "Nude on a Red Background"; the severity of Rudolph Schindler's untreated wooden beams intersecting with unadorned canvas-covered sliding door frames; and the comic yet sinister factory where Charlie Chaplin works in the film *Modern Times*. Regardless of their media, artists sensed the change, and filled their work with the sights, sounds, and even smells of industry, figuring the machine as central to the culture of the twentieth century.

While industrial machines popped a hundred years ago, information has emerged as the key figure for this new century. There are historical parallels between the emergence of the machine aesthetic in the first decades of the twentieth century and the nascent aesthetics of a digitized, unimodern culture in the twenty-first. The second half of the nineteenth century developed a market economy that produced and consumed machines. The early decades of the twentieth century saw artists, architects, and designers responding to this fever of material production by figuring the machine in their art, architecture, and design. The second half of the twentieth century, in turn, became an ever-accelerating feedback loop of information.

Thus, we should not be surprised that the past few years have seen the culture machine producing information-based art, architecture, design, and media; a digitized, interconnected society produces objects and systems that deal with software, databases, and the invisible flows of communications technology and computing algorithms. The great-grandchildren of those obsessed with Victoriana in the 1920s may look back with bemusement on their forebears' archaic tastes, but they are the ones flocking to modernist emporiums like the Conran Shop and Design Within Reach to purchase the highest expressions of the machine age at the very moment that the info-aesthetic is on us.

If we accept that in the era of the culture machine, information has popped to the forefront of our consciousness, then using figure/ground relationships can help us understand how unimodernism's electronic databases have transformed our expectation of stylistic "progress" and warped our cultural memory. When image, text, photo, graphic, and all manner of audiovisual records are available at the touch of a button anywhere in the unimodern wired world, the ordered progression through time is replaced by a blended presentness (what literary theorists would refer to as the replacement of the diachronic by the synchronic). So it is that there have been three or four distinct iterations of the punk aesthetic, since it first appeared in the 1970s. This churn is not so much an acceleration of nostalgia as a reworking of memory itself.

In the century and a half since the advent of the widespread reproduction of images, we have been through three distinct media regimes and are entering the fourth. The first was that of photography, dominating the second half of the nineteenth century. The photographic effect was utterly transformative, a mechanical technology that came to dominate the realm of images utterly. Photography affected the human understanding of everything from truth value, to the fluidity of time, to the very concept of what it is to be "creative." Picture-perfect, a moment in time, and an image at the click of a button moved from being marvels to clichés. The first half of the twentieth century was cinematic, with the movies creating the dream-worlds of the their viewers, their economies creating a new breed of human that we came to call stars, the logic of montage and the synergistic editing of moving image sequences reframing the very way we looked at life. As I have argued extensively here, the post–World War II era accepts the all-at-onceness of television and adapts to its fragmenting, zapping dramaturgy of the supermarket. We are the first generation, however, to have the computer as our culture machine, and much will depend on what we make of its networked capacities.

The Roman senator Cicero was famed for his mnemnotechnics, or the practice of memory. Twenty years after the fact, he could recite, word for word, speeches that he had heard on the senate floor. In a period before the wide availability of paper for taking notes, a trained memory was of inestimable value in governance and commerce.[3] Print transformed this situation, and by the Enlightenment, the arts of memory were already obsolete. If anything, the culture machine allows for even the outsourcing of our memories, with audio files, image banks, and video storage added to the archive. The effects of all of this storage go well beyond the memory of personal experience to encompass our memories of mediated experience as well. The universal database transforms the direct linkage between the object in time and the actual memory of that time. Television led the way, with children who grew up in the 1980s now "nostalgic" for their first viewings of *I Love Lucy*—a show that may well have predated their own parents' childhood. In other words, by the start of the twenty-first century, a uniform, temporally melded popular culture now exists that no longer needs stratification by decades.

Unimodern cultural production follows an arc first traced by Duchamp and his ready-mades. With *Fountain* (1917), Duchamp presented as sculpture a mass-produced urinal, turned upside down and mounted in a gallery. After Duchamp, it is the presentation of the object that defines that object's function within culture, with the shaping and molding of context come to the fore. This is not news, and in fact, those defining the differences between the high modern moment and what followed it hinged their definition precisely on this elevation of context to parity with the text itself. It has only been within the past decade that the combination of computers and communication networks has been robust enough to contribute to the creation of context. This context takes many forms, especially in relation to popular media, from the preplanned marketing of tie-ins from music CDs, television spin-offs, and lunch boxes, to the efflorescence of discursive communities generated by fans. In certain cases, these all combine to create something far more interesting than the backstory and more complicated than synergistic marketing. This is the "hypercontext," a dynamic, interlinked communicative community using networks to curate a series of shifting frames and content. The addition of greater levels of information to an object or system is not simply an additive process, it is a transformative one. It transforms objects by augmenting them and situating them in vastly larger hypercontexts, and when done in the proper spirit makes them stickier.

Being able to tell figure from ground in this environment of hypertrophied transtemporal bricolage becomes a vital part of negotiating the use of the culture machine. When the whole of popular culture from the last hundred years is finally brought under the disciplinarity of the universal database, it all becomes ground, and the refiguration of its parts becomes a veritable economic necessity. Those who are capable of refiguring in a way to attract an audience become fantastically powerful wealth generators—from hyperstylized director Quentin Tarantino to hyperintellectualized architect Rem Koolhaas, from Japanese Superflat artist Takashi Murakami to U.S. lifestyle guru Martha Stewart. How this figuration occurs, and how this process affects its meanings, defines the scale from Teflon to sticky. I have mentioned artists, designers, and directors here, but being able to flip between ground and figure is central to everyone's use of the culture machine. What we all, from world-famous designer to weekly blogger to occasional taker of digital snapshots, need is a catalog of strategies to help us understand what we download and contribute to what we upload. The ways that we figure words, sounds, images, and objects from the ground of information will define how and what we are able to produce with the culture machine. The key is to understand unimedia as the result and unimodernism as the aspiration.

Unimodern unimedia renders certain inherited categories less useful as we move into the future, but not without value as we analyze the present. How meaning manifests itself via the culture machine often links directly back into the specific histories of the individual media being simulated, and their traditions of authorship and reception. What follows here catalogs some of the strategies that these media have followed in this new era.

▼ SIDEBAR

The Soviet Man Who Fell to Earth

Of all the delightful thought experiments that theoretical physics has given birth to, from Erwin Schrödinger's cat to Richard Feynman's Brownian ratchet, my favorite is Albert Einstein's "twins paradox."[4] This story of two brothers explains the relativity of space and time. The first brother travels into space, while the other stays on Earth. The space farer is on a fast rocket and goes on a ten-year journey. When he returns home, though, he finds out that his brother has aged twenty years during his trip. This seeming paradox can be explained because of the way that traveling close to light speed shifts the vantage point for time. Science fiction squeezed this story for all it was worth, with the traveler losing decades, if not centuries, so that on his return, all that the spaceman knew had been consigned to a dusty past. I loved this story when I was younger in that way that certain children dream of the freedoms and powers of being orphaned. I filed this space/time paradox away for years, because I never expected to have to deal with it, at least until that unlikely moment that we achieved close to light speed travel. What physics proposed, however, history delivered, in the singular form of cosmonaut Sergei K. Krikalev.

Krikalev has spent more time in space than any other human: 803 days, 2 years combined, and counting.[5] In 1991, during his longest stint in orbit, this hero of the space age also came as close to experiencing the Einsteinian time paradox as one can without a close to light speed engine. When the Leningrad-born Krikalev left in May 1991, he was the USSR's best-known cosmonaut, launching safely from a Soviet base. When he returned to Earth ten months later, the USSR was no more, Leningrad had been renamed (or better, re-renamed) Saint Petersburg, and while he landed in the same "spot" from which he had launched, the base was no longer in the Soviet Union but rather in the newly independent state of Kazakhstan. In other words, Krikalev went up a Soviet and came back a Russian.[6]

Krikalev is something of a mascot, or perhaps better yet an icon, for this book. He is a power user of technology (what space farer is not?) and completely encapsulated by that same technology (for again, a human in the frigid vacuum of space cannot help but be so). He acts and is seen on both the local and global scale (he was selected to be the first cosmonaut to join the crew of a NASA space shuttle mission). He has been both a participant in history and an observer with a particularly stunning vista of it. One of the central ideas of this book is that in order to understand the impact of technology on culture, we need to regularly shift our perception of figure and ground. That is to say that we must look at the things that envelop us so completely that they become the background objects of our lives, and then force ourselves to concentrate on them as subjects—to figure them out of the ground. Accustomed to switching the ground of Earth for the figure of the sky, Krikalev's unique and true perception of our planet can serve as inspiration.

Words: The (Unending) Wonder of Hypertextuality

We are now so deeply entrenched in the era of word processing that we have forgotten how revolutionary the development of dynamic text was for the production of literature. That the culture machine can reformat your work while you are typing it, that you can grab chunks of it and rearrange them, that you can search for terms and replace them, and that the process of adding and editing is essentially one of unfinish—these are all the modes under which we work, so instantly ingrained that we have forgotten just how new they are.[7] When you add in hypertextuality, the ability to link and jump from one section of a text to another, or from one text to an entirely different one, you have one of the defining qualities of the unimodern culture machine.

Hypertext showed the way by making the link integral to the construction of the meaning. The creation of meaning via juxtaposition is ancient, of course, but the modern era's refinement of collage in still images and then montage in the cinema elevated the status of the meanings produced through these processes. The televisual era introduced a randomness to the juxtapositions. If Soviet filmmaker Eisenstein's dialectical montage was about the deliberate production of effect through cinematic editing, channel zapping on television was closer to the experimental "cut-up" fiction of Brion Gysin and William Burroughs in the late 1950s. Gysin, a painter, and Burroughs, a novelist, created texts and then literally cut them up into pieces, reassembling the fragments at random, giving up a large measure, though not by any means all, authorial control.[8]

The earliest attempts at hypertext tried to marry the randomness of the cut-up technique to a restricted universe of potential connections, thereby establishing the technoliterary equivalent of a forced card in magic. You had choice as a user/reader, but your choices and paths were often predetermined by the author. The advent of the World Wide Web broke open these closed text worlds, creating the freedom to jump around with "real" randomness. One of the earliest net.art text pieces understood the new environment perfectly, linking every word on a Web page to a domain that contained that word—a far more inventive concept in 1996, when there were thousands rather than billions of pages in ether space.[9] What is new in the world is that text more and more becomes something that is linked to anything, words become the building blocks of augmentations, the whole world develops labels like those at museum exhibitions, and each label links to another one describing, advertising, or commenting on another text, another image, another object. The hyperlinking that starts with text as far back as the 1940s' experiments of Bush and Turing becomes the default mode of figuring "meaning" in the world. What happens with text moves on to sounds, then to images, and finally to physical objects.

Sounds: Mix and Mash

For the culture machine, it is as though everything that happens in the realm of the visual happens years before, first with text and then with sound. Sound is cheaper and easier to store, manipulate, and upload than images, and so it has been that digital technologies have transformed not only the media that the music arrives on but also the very aesthetics and content of that music. The shift from analog to digital is about much more than the shift from vinyl albums to CDs, and then to free-floating file sharing. The proliferation of cheap synthesizers and editing suites enabled by digital technologies spread this meme to musicians and producers worldwide, and the music itself began to change. By the time the culture machine eventually simulated and subsumed these sound-generating and sound-organizing modalities, an entire generation of listeners were creating sampled, remixed, digitally processed, digitally accessed music. From the now-quaint "You've Got Mail!" AOL voice-coder greeting, to the advent of audible interfaces and game soundscapes, to the popularity of pop snippets as personal audio identifications in cell phone ringtones, there has been a proliferation of audio cues within work, play, and mobile environments.

The unimodern soundscape owes a huge debt to hip-hop culture. The origins of hip-hop are to be found in the analog arena. In the 1970s, disc jockeys in the Bronx cut back and forth between turntables with vinyl records on them, mastering their ability to "drop samples" and use the turntables themselves to generate new sounds—the ubiquitous "scratching" of that era. But within a decade, the culture machine started to absorb and simulate these analog techniques, and the digital sample became the music's building block, and remixing became the aesthetic strategy of choice. Hip-hop and high tech are inextricably bound together, offering a sterling example of the street finding its own uses for technology.

The now-defunct music file-sharing service Napster makes for a good case study of sound's pioneering media status. People had been using electronic networks to trade and collect songs, and even whole albums, long before the advent of the Web, but it was only after a Northeastern University student in Boston named Shawn Fanning created a decentralized, easy-to-use system for organizing, making available, and downloading music that the peer-to-peer phenomenon really took off. Napster made it much simpler for network users to look through other people's digitized music libraries, download what interested them, and open their own libraries so that others might do the same. The effect on the music industry, the intellectual property issues, the legal prosecution of downloaders, and the impassioned positions taken by artists, music executives, and fans have generated megabytes of commentary, and the business history of Napster has been covered in legal journals and been the subject of a shelf's worth of books.[10] In the end, though, what Napster managed to do was effect a qualitative difference in the ways in which people thought about the network and the Web. The network was now open for sharing to the wired masses rather than being an exclusive preserve of dedicated hackers. The meme of sharing, regardless of the issues of copyright, became embedded deeper and deeper into the habits of users, preparing the way for YouTube and other participatory sites.

One place to see the hip-hop collage aesthetic collide with post-Napster file sharing is the phenomenon of the mash-up. Mash-ups meld two or more recordings into a new entity, most famously done by Danger Mouse when he mashed the Beatles'

White Album, a defining work of the 1960s' rock era, with Jay Z's rap epic *The Black Album* (2003) to create *The Grey Album* (2005). The result was widely distributed because of the Web, file sharing, and the proliferation of sound and image editing tools. The ability to download vast archives of music, whether accessed legally or (more likely) illegally, allowed for an explosion of mash-ups. The fad, for it was a fad, eventually died down, as Web-driven phenomena frequently do, but the mash-ups were proof that huge audiences were playing with their culture machines, mixing, matching, pasting, and then getting that unimodern material out into the unimodern world.

Images: WYSIWYG, or What You See Is What You Get

What happened in text and sound inevitably spread to the realm of the image. The explosion of cultural production that mash-ups reflect has in turn transformed our understanding of the meanings of words like "print" and "publish." We print much more than text these days. The first major shift came in the era of desktop publishing. In the digital realm, text and image are just strings of ones and zeros, indistinguishable as information, and made manifest only by the medium in which they are eventually released. So an image could be fluid in an animation, printed on paper as a screen, encompassed in a resizable window with surrounding text, or blended in a graphic with those same typographic elements, which could themselves be animated as a motion graphic.

The designer Bruce Mau refers to "PostScript World" when he discusses the radical transformation that the culture machine brings to our visual environment. With the development of "page description languages" like PostScript from Adobe Systems, there is "no longer any distinction between text and non-text, image and non-image." Surfaces are "now described in one language. Everything is now image."[11] PostScript World announced itself with the desktop publishing phenomenon, in which the image on the monitor looks like the page that the printer will produce, and vice versa.[12] This was the software/hardware combination that brought us the acronym WYSIWYG, for What You See Is What You Get.

The previously independent realms of word and image were now brought together under the sovereignty of PostScript World. What had once been the realm of obscure pasteup artists, burly press operators, and black-clad design gurus became a commonplace at every office worldwide. In 1970, only the most design savvy knew what people meant by the term "font"; three decades later, second graders talk about their favorite letterforms with a passion formerly reserved for toy trains and paper dolls. When images and words are both expressed in the same code, the distinction between them erodes, and people speak with images and paint with type. As the PostScript World came to embrace the mutability of PhotoShop as well as the development of animation and motion softwares like embedded digital video, centuries-old distinctions between media forms dissolved in turn and created unimodern unimedia, the digital soup that the networked culture machine pumps worldwide.

Dynamic Media: From Microcinema to Macrotelevision

Twenty-first-century unimodernism enables the utter promiscuity of images. I use the term "enables" in both its technical and pop-psychological senses. Advances in technology allow moving images to break free of their respective media and cross-pollinate device to device. Here again, though, the quantitative shift has brought about a qualitative change. When movies show up on cell phones, sitcoms are streamed to desktop monitors, motion graphic festivals take place in cinemas, and viral Internet videos are ported to massive flat-screen televisions, the idea of "medium specificity" is not so much challenged as destroyed. Notions of "appropriate" scale, run time, and even content are now all up for grabs. The image goes wherever it wants to and is wanted. The image shrinks and expands, from digital video on personal digital assistants to environmental moving graphics as large as the city blocks they are already blanketing in New York's Times Square, London's Piccadilly Circus, and Tokyo's Ginza.

The substrate of these moving images can be screens or surfaces to be projected on, and they can function as art, but more likely serve as commerce or dynamic advertisement. When you add in interactivity, they can become everything from responsive visual environments to video games. In the era of the digital culture machine, the moving image, whether augmented by sound or not, becomes even more dominant than it was during the twentieth century. In terms of scale, I categorize this as everything from microcinema to macrotelevision, with screen culture existing everywhere from handheld devices to animated, building-sized supergraphics.

Broadband technologies vastly increased the uploading and downloading of images, and especially moving images, worldwide. This access created a newly ubiquitous archive of movies, motion graphics, large-scale networked simulations, and visually rich interactive gaming environments. Hyperlinking moves beyond text, sampling enters the realm of the visual, and visual mash-ups become twenty-first-century extensions of the collage and montage aesthetics that dominated the previous hundred years. Formally, the twenty-first-century moving image is defiantly heterogeneous. The dominant style stretching from microcinema to macrotelevison is multilayered and processed.[13] The default expectation is that the image will be hybridized. Referring back to classic cinematographic style, but also embracing the raw pixilation of amateur video, screens are overrun with graphics that can trace their lineage in any number of directions, from film titles to animation to pop art to infoviz. The word has come roaring back, with textual emendations popping up, and the image surrounded by alphanumeric data about everything from the rise and fall of markets to upcoming programming. Transitions between elements are ever more fluid as the now-classic cut finds itself supplanted by the ever more ubiquitous fade, with elements coexisting in the image's now-expanding space/time.

One little noticed transformation has been the way that the culture machine reduces the unity of the "cinematic." Computer screens so frequently contain textual elements, interfaces, and multiple other still and moving image windows even as it is used to display "movies." The computer's multiple image, text-rich interface has had an impact on its predecessor media. This influence ranges from the cluttered daily feed on CNN's *Headline News*, to the split-screen approach that Mike Figgis used in his pioneering film *Timecode* (2000), which told four intersecting stories continuously, in real time, in four quadrants of the screen, to the complex navigation of DVD menus. The impact of the culture machine on the cinematic and televisual deserves, and has generated, its own libraries of analyses.

If these are the new "styles" of moving image media, what about their contents, and the stories they tell in particular? Stories used to be the figures, brought into focus out of the grounds of daily life. All stories are in effect moral tales, instructions on how to live, act, and accept our fates. But in the twentieth century, we were all suffused in narrative, floating in a bath of professionalized storytelling, in the pages of magazines (both the upscale "slicks" and cheap "pulps"), in comics as well as hardcover and paperback books, on the radio, in the cinema, and on our televisions. Now that narrative surrounds us, it has become the new ground. When the actor Bruce Willis was confronted at a press event about the amiably shambling incoherence of his recent digital-effects-driven film, he laughed and let the scribes know that nobody cares about the story anymore.[14] While we do not generally expect our action heroes to moonlight as narratologists, Willis's observation was at least partially accurate. We simply have so much narrative surrounding us that it is usually not even necessary to recount it. Like sampling within contemporary music and so much of our endlessly referential advertising culture, a nod to an established and overflowing narrative tradition is sufficient.

The movement toward a referential rather than developmental narrative strategy is an outgrowth of the sheer plentitude of narrative, figured most emblematically by the glowingly accessible archive of everything in the era of microcinema and macrotelevision. The Web's 24/7 access makes the video store as archaic as the repertory theater. But let us not forget that the art of cinema, and film culture itself, was healthier in the period of the rep theater than it is now. The very proliferation and ubiquitizing of narrative, even the highest-quality narrative, can have the paradoxical effect of making it seem that much less important—Willis's sense that "nobody cares."

This is an unintended consequence of unimodern communications technology on the freedom of access and even discourse. The underground circuit of mimeographed manuscripts known as samizdat circulated throughout Eastern Europe and the Soviet Union during the 1970s. After the events of 1989 through 1991, however, literary markets emerged in these countries, and something both ineffable and significant was lost. The Web offers a marvelous explosion of access, but the law of unintended consequences could usher in a world in which anything can be obtained, but nothing is special.[15] Whether the pluralistic possibilities that digital video connoisseurship offers—either off disc or online—trumps the loss of the communal viewing

and discourse that the cinema engendered remains to be seen, though the potentialities for a serious discussion of the range from microcinema to macrotelevision in networked environments is ever growing.

What one can talk about with certainty are the unintended effects of an ever-filling pool of information. The first is the accessibility of style cannibalization. The interpenetration of communication networks, combined with their increasing memory, means that ever more detailed records of the past's art, music, and design are available for consultation, inspiration, and fandom. At its best, this archive fever can produce a deeply textured, historically informed collage aesthetic. Yet just as easily, you can see the development of referential churn as historical styles and allusions are recycled on an ever-shrinking cycle. The work that went into older versions of historical revivalism (from the Renaissance forward) is now beside the point. It took months of research by Yves Saint Laurent to do his 1940s' collection in the 1970s, and the duration of the research period itself may have allowed Saint Laurent the capacity to reflect the 1970s as well as the 1940s.[16] Deep reflection is by no means a default setting given the immediacy of the culture machine's archives. I would posit that these mimicry issues will always be with us, but that as the networked archives densify and become ever richer, artists, writers, designers, and users will learn to take advantage of the new modes of access, storage, and manipulation, and develop ever-stickier media.[17]

▼ SIDEBAR

Here We Are Now, Entertain Us

"This story is about truth, beauty, freedom; but above all things, this story is about love." So begins Baz Luhrmann's *Moulin*

Rouge (2001), which for all its references to the past (and even to love) is the first great studio movie about twenty-first-century unimodern culture. What Luhrmann, who had earlier directed a hip-hop version of *Romeo and Juliet*, and came to acclaim with a film about ballroom dancing, created was the first major open-source opera—even if he had to pay for every bit of it with studio money. That his fin de siècle Paris is spectacularly phony does no disservice to the layers of so-called original material he so deftly reworked.

First, there is the enduring myth of the avant-garde, of artists in the garrets willing to die for love, and of the madness of creation and the tragedy of commerce. These were the working concept for French novelist Henri Murger when he sketched out his *Scènes de la vie Bohème* in 1846. Murger took his title from the poor Czech immigrants (from the principality of Bohemia) who populated the neighborhood, for the artists, musicians, and writers who recently had found refuge in the local lofts and attics. Fifty years later, the Italian composer Giacomo Puccini premiered an opera based on this pseudodocumentary material. First performed in Turin, Puccini's *La Bohème* has since become what the *New York Times* claims is the "world's most popular opera—a rite of passage for generations of those who would be besotted by the tale of bohemians (now a creative class rather than an ethnic group) who make love and art, and who suffer and die.[18] After all, *La Bohème* is both romantic and Romantic.

The opera was loosely adapted in 2001 by an Australian film director using U.S. studio money. Luhrmann's *Moulin Rouge* tells the same story as the opera and book, and the literally endless variations on these themes from the Lillian Gish 1926 silent melodrama through Jonathan Larson's 1990s' musical *Rent*, which switched the locale to New York's East Village in the era of AIDS.[19] What distinguishes Luhrmann's film is the way that it deploys this myth from the nineteenth century to analyze, dissect, and recombine the popular culture of the twentieth. The film provides the antithesis of purity, proffering instead a spectacular hodgepodge, a mélange, and a remixer's delight. Luhrmann's insight is that the kind of devoted and minute dissection of twentieth-century popular culture by decade and style—1920s' fashion, 1950s' music, and 1980s' hair—was disappearing through the passage of time and under the weight of the ever-growing archive. The references in his film move effortlessly from place to place and decade to decade—Parisian streetscapes of the 1880s, an Argentine tango straight out of 1933, and a Bollywood-style number inspired by contemporary Mumbai. But more than anything else, it is the aural environment that is most open to this playful and powerful remixing. Although Luhrmann cannot resist Jacques Offenbach's "Can-Can" ("Orpheus in the Underworld," 1876), the rest of the music is a medley of pop hits stretching over decades. The haunting 1950s' ballad "Nature Boy," made famous by Nat King Cole, exists in the same sonic collage as the Beatles's 1960s' hit "All You Need Is Love," David Bowie's decadent "Diamond Dogs" from the 1970s, and Nirvana's grunge anthem "Smells Like Teen Spirit" from the 1990s. These tracks and more are all deployed not as "signifiers" of their respective eras but instead as a "sound track of our lives." Rather than the kind of detailed specificities that we came to associate with "oldies" stations versus "new wave nights" and retro-greaser punctiliousness about the "origins" of rock and roll, *Moulin Rouge* takes its cue from Kurt Cobain's most famous lyric: "Here we are now, entertain us."

As opposed to simply numbing us with this flurry of reference and citation, however, Luhrmann in fact provokes us to a new level of understanding about our historical positioning in relation to the ubiquity of mediated entertainment. From the perspective of the denizens of the *Moulin Rouge*, the twentieth century is now a unity in our heads. It is the place and time where Madonna and Frank Sinatra, Tin Pan Alley and disco, and silent cinema and MTV all come together as the state of entertainment. Just as few beyond historians, classicists, and devoted amateur Hellenophiles could tell you the difference between an Athenian bust produced in 390 BCE and that from 310 BCE–it is all fourth-century glories of Greece to the vast majority of us–so too is the century just past treated by Luhrmann.

More than the vertiginous pleasure that this intelligent spectacle offers us, there is also a lesson in the importance of access to what have become the constituent parts of our culture. Luhrmann could not have reinvented the musical without access to songs that defined love for him as well as for many of the rest of us. In the defining moment of the film, the young writer woos the beautiful courtesan with song. In earlier musicals, this ballad might have been a new one, hoping to join the pantheon of love songs, but here it is the pantheon itself that springs from his lips. The "Elephant Love Medley" (so named because he sings it in the courtesan's pachyderm-shaped and themed boudoir) comprises thirteen popular love songs intertwined and remixed to draw together figures as diverse as Sinatra and Kiss.[20] Luhrmann, of course, is able to do this because Fox Studios had the budget to pay for the performance rights to all these songs. But his very success and the way that he proves how these media products have come to shape us so perfectly also makes the case for allowing other artists access to the archive. What he likewise proves is that durable and significant creative work is being done in the unimodern era of cut and paste. *Moulin Rouge* is a far more potent artifact of the digital age than any amount of special effects wizardry in a science-fiction or fantasy film.

Objects and Spaces: WYMIWYM, or What You Model Is What You Manufacture

From words to sounds to pictures to moving images, the networked computer has transformed the production of culture. The next new thing that is in fact already here is the "printing" of objects. The Postscript World of image/text printing has become part of an even larger system of computer fabrication, or "fabbing," in which what was once restricted to two dimensions is extruded in three. WYSIWYG, What You See Is What You Get, is being followed by an era of what I call WYMIWYM, for What You Model Is What You Manufacture. Just as WYSIWYG allowed new freedoms to graphic designers and two-dimensional image makers, the WYMIWYM era of computing allows architecture and industrial design to play with form and iteration, and make complex extant forms easier to manufacture profitably. In other words, what the computer did to the flat, two-dimensional fields of painting, photography, and graphics is now happening in the three-dimensional realms of sculpture, industrial design, and architecture, as artists, designers, and architects develop forms on the computer, and then fabricate them with three-dimensional printers.[21]

An architect like Greg Lynn can use three-dimensional printing to do everything from creating maquettes, or small-scale models, of buildings to making prototypes for designs for a line of flatware commissioned by the Italian design manufacturer Alessi. When the fabbing specialists at the design collaborative Machine Histories worked with artist Pae White to create a complex bedframe for an exhibition, they worked with solid Corian, usually a surfacing material in kitchens and bathrooms. The object, titled "widow of a king bedframe" (2006), was so intricately worked by Machine Histories's unique tool paths that it felt airier than one would ever expect a headboard to be. The deft carving and intricate detailing went beyond what handwork could have accomplished, and serves as a reminder that expertise in three-dimensional fabrication will indeed bring on a new material culture for the twenty-first century. This is all the more true because art, design, and architecture students are getting exposure to 3-D modeling tools along with large-scale 3-D printers, extruders, and other computer-aided manufacturing in school now, and you can bet that they will fill their own studios and ateliers in the future with the smaller, cheaper 3-D printers that are already in development by the manufacturers.

These WYMIWYM objects obviously figure informationalism in their production process, but as they themselves become linked into larger networks, through the incorporation of sensors, transmitters, and augmentation, they begin to attain autonomy. From mute objects and closed spaces, they become nodes in the network, aware of their place and time, and capable of communication from the minimal to the maximal. The incorporation of radio frequency identification devices (RFIDs) and microcontrollers into formerly quotidian objects enlivens them in an almost magical way. Like the animated brooms in Walt Disney's *Fantasia* that come alive when Mickey Mouse accidentally enchants them as the Sorcerer's Apprentice, there is a glamour, in its magical rather than fashionable sense, inherent in these new, augmented objects and spaces.

The explosion of WYMIWYM objects and spaces will bring about an efflorescence of style, just as WYSIWYG publishing did. Much of it will be excruciatingly bad, worse even than bad desktop publishing because it will have more dimensions to fill with its awfulness, but this is to be expected and embraced. Much that is wonderful will also be discovered, and perhaps some of what makes us wince will eventually earn at least grudging respect for its exuberance. But the ability to follow a program, in the architectural sense of an overarching vision, that the WYMIWYM era allows can engender the opposite problem from that of too much unstudied pluralism: it can also allow for the figuration of information in too perfect a form.

Karl Kraus, a Viennese modernist in the early 1900s, once complained that *art nouveau* living spaces were so fully integrated that they allowed their inhabitants no "running room" for the imagination. In the emerging clusters of entertainment design and experience design we see the resurgence of the totalizing impulse. The Disney World model of complete design integration from food to signage to people mover to thrill ride to collectible souvenir moves centrifugally outward from its Orlando home, becoming the de facto model for new experiences within entertainment capitalism. One factor contributing to the rise of entertainment and experience design is the computer itself, which allows for an unprecedented merging of design disciplinarities along with a sharing of communication and information across design groups, participating companies, and geographic space.

The impact of these intersecting design and technology schema are to be found everywhere from the branding overkill of themed resorts like Paris, Las Vegas to Jean Nouvel's seamlessly integrated galleries of indigenous art at the Musée du quai Branly in Paris, France. Here, as in so many other hyperdesigned spaces around the world, interface and object, building and Web presence, as well as commodity and brand identity all swirl together in unimodern, digitally enabled Postscript documents and WYMIWYM environments.

The figuring of informationalism into form has been our preoccupation in this section, and these forms—as words, sounds, images, objects, and even spaces—serve as semantic building blocks for the syntactic ways with which we will "speak" with these media. The secret war between downloading and uploading is predicated on the idea that the message and its meaningfulness need our full attention as well.

Play: Modders and Other Do-it-yourself Pleasures

People's willingness to embrace unfinish differs by age and class—that is to say, by who can afford it in the first place. Sometimes the adults who design systems can forget how much younger users are invested in finding ways to fill their downtime. Television, music, and video games can all be seen as preemployment time fillers for adolescents, and even those self-styled "rejuveniles" who are choosing not to abandon the games and pastimes of their youth.[22] But those with the desire and access to the culture machine can kick-start their own do-it-yourself (DIY) movements. There are deep desires to categorize and annotate one's own life as well as the lives of one's friends and community. This moment is not about professional narratives so much as the development of new tools to create letters, diaries, photo collages, and home movies.[23]

At its best, these DIY archives transform lived experiences not into commodities sold back to us but instead as realized memory traces that we construct ourselves and communicate to communities of interest. These actions indicate that the desire for the personal rather than the professional archive is ever expanding. From the mimeograph machine, to the advent of videotape, to fax technologies, to public access cable television, each new communication technology brings with it a new potential for participation. Think of the copier machine, which was a huge boon to the punk era, when fans produced zines (the small magazines and fan letters that were created out of a sense that *Rolling Stone* and the other major magazines would never "get" punk). The computer has encouraged the growth of new forms of DIY, hacktivist, and even craftivist culture.

Take, for example, the crafting Web site etsy.com. It is composed in almost equal measure of three apparently unconnected concepts: an enthusiasm for alternatives to mass-produced objects, e-commerce capacities inspired by the success of eBay and Amazon, and the gestalt of a summer craft fair in Vermont. Etsy has grown by attracting a young, primarily female user base that is interested in making, selling, and buying handmade objects. The site's rhetoric and design schema are carefully considered to attract just such a demographic, of course, but there is also a sense that etsy would and could not exist without the authentic excitement of its users for a space that could not have ranged as widely before the Net provided the affordances for such a community. One of the interesting evolutions of the site has been the growth of the "buy local" option that allows members to develop place-based networks as well as national and international ones. Etsy's users want to create a different relationship to their material positions, carve out a space in which makers can communicate and trade, and build what essentially become microeconomic relationships that are personal rather than corporate.

MAKE, a magazine, Web site, PBS television series, book line, and succession of public "Faires" takes DIY concepts and makes them available in an ever-expanding set of interrelated media. Mark Frauenfelder, *MAKE* magazine's founding editor in chief, brought a great deal of credibility to his publishers when he proposed a concept for engaging with the remarkable explosion of objects made by and with the culture machine. Frauenfelder had been involved the cyberpunk print fanzine *Boing Boing*. After migrating to the Web as boingboing.net, it grew into a huge "directory of wonderful things," as *Boing Boing* says in its masthead. The site's studied eccentricity, the indefatigable energy of the four principle bloggers, and the bloggers' worldwide network of interesting collaborators exposed both Frauenfelder and his boingboing.net readers to everything from long and serious discussions about culture jamming to a prototype for a polite umbrella that contracts to avoid poking other people in the eye.

Frauenfelder's next move was to create a separate entity to concentrate on the making of this kind of culture—a twenty-first-century hybrid of *Popular Mechanics* and *Martha Stewart Living*. *MAKE* magazine's first issue came out in 2004, and since then it has covered everything from crafting interactive fashion to creating personal lighter-than-air dirigible flying robots.

The emphasis is on producing new and networked objects, and the response was strong enough that Frauenfelder and his coworkers decided that they could expand into producing live events to bring together their community, offering demonstrations and workshops, and growing the number of people interested in these new DIY phenomena. The resulting events, called MAKER Faires, drew from other communities, like the DIYers who have been such a huge part of the Burning Man festival in the Nevada desert, and became social spaces that blended consumption and production, fan and maker, and online interaction with real-life excitement. The point here is less the commercial success and long-term viability of the etsy and *MAKE* DIY communities than the ways in which their very existence points toward a future of blended real and virtual communities devoted to the material production of culture along with its integration into more open spaces of commerce, trade, and exchange.

The ease with which people can build a like-minded community combines with the ability to share component software as well as reports on process and results. There are knitters using networks to expand their discussions about their craft, the open-source software and hacker communities, and then interesting hybrids like "modders," as those doing electronic modifications call themselves. These people take mass-produced objects and change or modify them in a way to "make personal" the products of an advanced technological society. The sheer amount of craft and obsession that went into the process of remaking an iPod out of hardwood, including a working jog wheel, boggles the mind, but it is a quintessential mod.[24] This is a physicalized metaphor for remix culture—taking something, adding one's own spin, and putting it back out into the world (with mods, it is often just pictures of the object and its production process). But the more bit-driven realms of remix culture differ in that the remixes are then sent back out into the world to be remixed again themselves in a recursive and ever-unfinished loop.

Gaming: Ludic Stickiness

Certain media are either emboldened or diminished by the expectation that "in the future" they will become somehow that much more than they already are. Games, for example—like comic books, or "graphic novels" as the recent rebrand would have it—have long been in just such a situation. Although there is no area in which the computer as culture machine has come to so dominate, games are still seen in many quadrants as forever on the verge of crossing over into a realm of deeper meaning and greater cultural impact. Part of this tentative embrace of the gaming medium is that the worlds that games create have steep entry costs—not so much in terms of money or even access, but rather temporally. To master the skills required to play proficiently enough to enjoy gaming itself is merely the first investment of time. The next, and perhaps most serious in terms of this discussion, is the time needed to simply explore the game space sufficiently to see it as more than a fragment. This can be ten, twenty, forty, or even eighty hours of commitment. That strikes committed gamers as a fine value for the money invested in the purchase of the game, but the sheer time demanded tends to deter the uncommitted or "casual" gamer, much less the bystander who might be interested in the experience, yet cannot justify such an expenditure of time. In this, gaming is quite different from the cinema, where a 90- to 150-minute commitment is all it takes to be part of the "experience."

One way to understand this divergence is to realize that for all their narrative conventions, games are not best understood as interactive stories. To get a feel for what matters in gaming it is worth revisiting their earliest history, before gaming's visuals came to rival the realism of cinema and television. Although there was a tic-tac-toe game and a tennis simulator in the 1950s, it was really Spacewar!—developed by students at MIT in 1962 for their own amusement—that stands as the urtext of gaming. With two armed ships shooting at each other while spiraling down a gravity well, Spacewar! established a few conventions of gaming that remain powerful today. These include conflict, time limits, and graphic interaction.

The game itself was a useful way to gauge the speed and accuracy of the Digital Equipment Corporation's PDP minicomputers, and the company began to ship later units with the game in the core memory. This ensured an ever-growing group of users, who would go on to create later pioneering games for arcades and the growing home market, including Pong, Space Invaders, and Pac-Man. Arcades, consoles, computers, and handhelds—these and more were the material substrate of gaming. Over the years, designers have configured their games for single players, for a few players arranged around a television, or for millions spread out worldwide on the Net in massive-multiplayer configurations. What has not changed, no matter what the era or configuration, is the importance and specifics of game play.[25]

There is no question that games have become a fantastically successful part of the culture machine's impact. For their players, there is no denying that gaming brings a level of enjoyment equaling sport and a level of immersion that comes to rival architecture itself. The power of gaming to involve the committed, then, is hardly worth discussing. The longer-term issue is whether those gamers will in turn effect the culture as a whole or whether the ludic experience will be restricted to its own, hermetically sealed world. As haptic and other interfaces become more widespread in the wake of Nintendo's success with the Wii system, whether or not those casual players become more involved with other forms of game play remains to be seen.

Two other arguments tangential to play itself have dominated discussions about gaming. The first is the effect of violence in the game space on violence in the real world, and the second is about the influence of gaming's twitch culture on cognition. The first is an argument about content for the most part, and while it has a great appeal for parents concerned about exterior influences as well as the politicians who cater to these voters' concerns, this is a contention that holds less and less interest as "shooters" become more and more a specific genre of game rather than an overarching category. The neuroscience and cognitive science studies on gaming are still coming in, and critics, depending on their preconceptions, divide into two camps, either bemoaning the splintering of attention that video games bring in their wake, or lauding the response time and multitasking skills that games engender in their most avid players. These are all serious issues, spanning the range from the sociological impact of repetitive actions to the neural conditioning that distinguishes gaming from other media. In the context of the assertions offered in the rest of this book, however, I would say that the pressing issue is whether individual games or games as systems can accrete in such a way as to create what one could call ludic stickiness.

One game that was indeed sticky involved players running around a huge and unconventional map of the world, working together to deploy resources and innovative technology to make not just their team but rather the whole globe a better place. More than a generation ago, the polymath futurist and designer R. Buckminster Fuller (of geodesic dome fame) proposed this multiplayer "design science process for arriving at economic, technological and social insights pertinent to humanity's future envolvement [*sic*, a signature Fuller neologism] aboard our planet Earth." Originally called the "great logistics game" and then the "world peace game," it was best known simply as the "World Game." Inspired in part by the war gaming that planners engaged in to prepare for the hot battlefields of World War II and the colder, yet protracted conflicts with the Soviet Union that followed, the World Game was a revamping of these strategies to think about how best to use resources to ensure planetary happiness.

Often laid out on the unfolded polyhedron of Fuller's own Dymaxion map, the game used a synergistic rather than competitive play strategy to determine ways to best harness the natural resources of the planet. Fuller's map gives a better sense of the relative sizes of the continents than the usual Mercator projections, and even more subversively does not have a natural "up" or "down" that de-privileges people's usual expectations of maps and the sense of space that they project. Fuller maintained that the goal was to "make the world work, for 100% of humanity, in the shortest possible time, through spontaneous cooperation, without ecological offense or the disadvantage of anyone." The World Game was a product of postscarcity thinking and 1960s' utopianism, played without benefit of networks and computer simulations, but its essential message—that humans working together have the potential to craft a better world—resonates, and more than ever looks like a prototype for the networked effects of simulation and participation.[26]

Running Room or Play Space?

Simulation and participation drive everything from figuring information to the fabbing of WYMIWYM objects; they make possible the mixing and mashing of open-source sound and imagescapes; and they shape the ways that we work as well as the ways that we play. It is my hope that the detailed listing of all these manifestations of the computer as culture machine in aggregate proves the existence of the unimodern unimedia posited at the start of this chapter. In keeping with the spirit of this project, I hope to not simply identify unimodernism but to point toward ways in which its unimodern unimedia might deepen meaning and engagement with the world, art, and each other. What we need to confront is the explosion of information that computer networks engender.

Understanding the changes wrought by computer-inflected technologies point to the huge difference between processing data and designing its output. This conceptual clarity will also help us to categorize what kind of culture we are actually constructing in the twenty-first century. If we divide the last century into early modern, high modern, and postmodern strands (roughly 1900–1919, 1919–1973, and 1973–2001, respectively), the culture machine's ubiquity has braided all three (and more) into unimodernism. The twenty-first-century culture machine's modernisms exist simultaneously in an ever-present database, ready to be deployed or redeployed in the cultural equivalent of just-in-time production.

The single most important issue is to ensure that the uniformity of substrate that the computer brings to culture does not produce a stultifying sameness of content. To do so, it is worth revisiting Karl Kraus's concept of running room. In the original German, the word is *Spielraum*, the roots being *Spiel*, or "play," and *Raum*, or "space." So whether running room or play space, the concept brings with it a sense of exploration, imagination, and engagement with the unexpected. The sheer productive capacity of unimodern unimedia can and should be able to carve out this *Spielraum*. Running room is different from the touted benefits of diversity, however, because diversity is often another way to describe the offerings in a bazaar. If the diversity that is being offered is simply in the realm of consumption, it remains just that: consumption.

The play space I am discussing will be located within twenty-first-century capitalism, but it has to offer the choice *not* to buy and especially the option to *make*. That is one reason that the open-source community is so crucial to the future of running room. Free culture as a gift exchange offers a real challenge to the inherited affordances of market economies. The generosity of online communities serves as a way to access the powers of the always already available archive of the unimodern culture machine without falling prey to the notion that the market defines everything and that the imagination must be tied to its precepts.

We have already seen how unimodern unimedia has exploded access and content in our cultural archives. This expansion has in turn lead to more opportunities for collaborative multi-authorship. This kind of unsigned multiple creatorship is reminiscent of the Greek myths and the Great Wall of China. Both the myths and the wall took centuries to build, and thousands of people contributed to their effort. We build multiple author works as well, but now we call them Linux, Wikipedia, Flickr, de.lic.ious, and communal bogs. These are the cultural forms that show us a future in which we could all potentially contribute to the creation of things and systems vastly larger than ourselves. This has frequently been the effect of religious devotion, of course, and those who have been to a barn raising have experienced similar kinds of emotions.

Earlier we saw how the memes of simulation and participation competed as well as built on each other: simulation enabled functionality, and participation brought that functionality to ever-more people. This was the promise of computing, and the cultures it has engendered differ radically from those we inherited from a half century of television viewing. The previous regime offered and continues to beguile us with an ever-increasing plentitude of narrative entertainment (again, it does not matter whether that entertainment was called a situation comedy, the nightly news, a shopping channel, or a reality show—it was and is all entertainment); it creates habits of mind and modes of consumption that lead to cultural diabetes.

The development of ever more complicated and intertwined systems of delivery, the downloading syndrome, can lead to a proliferation of meaning-lite, if not outright meaningless, content. That is why, in a book that set out to celebrate the best of the culture machine and its products, there is an underlying fear of unexplored avenues that will shut down in the face of an inexorable yet barely perceptible pressure to do less rather than more. This book shuttles between the past, present, and future, and one of the fears it deals with is the concern that no matter what they want, people may end up getting a machine that emulates their televisions, but with a cell phone and credit card shopping grafted on to it.

Combine stickiness and unfinish, however, and what you create are ever-enfolding and expanding interconnections of hypercontexts. Those who want to do new work with the culture machine must ensure running room for the imagination as well as playful space for mindful downloading and meaningful uploading. This is the unimodern dream—less grand than its predecessors perhaps, but no less worthy.

CHAPTER FOUR

WEB n.Ø

▶ notes: pp. 188–189

Environmental Impact Report

The mid-twentieth-century critic Lionel Trilling understood literature to be "the human activity that takes the fullest and most precise account of variousness, possibility, complexity, and difficulty." We have been looking at the culture machine to see whether there is evidence of twenty-first-century variousness, possibility, complexity, and difficulty. Sticky media, unfinish, mindful downloading, and meaningful uploading matter at this moment because the general public has come to see the networks not just as stable technologies like radio or television but instead as a system that evolves both quickly and, more important, radically over time. Thus it is that just a few years after the first dot-com bust, people began to speak of a "second" version of the Internet, which they called Web 2.0. This happened at the same time that the technology sector started to regain its footing, and the emergent, highly social nature of Web 2.0 reignited an investing frenzy. Labeling these social media phenomena as 2.0 may be premature, however—akin to labeling 1920s' Dada as postmodern because it emerged a decade after cubism. Rather than Web 2.0, it may be more useful to think in terms of evolution, looking to mathematics for the vocabulary to describe a continued evolution of the Web to the nth version, or Web n.0 for short.

The Web has now been around long enough to think through what we might mean by "developing" it. Any time that a major land development is proposed, part of the process of determining whether to proceed is to issue an environmental impact report, or an EIR as they are known. EIRs measure the effects that any building would have on the quality of the human and natural environment surrounding it. EIRs discuss a range of alternatives and strategies, and assess the impact of each combination of factors. Assessing these factors is always important, but it becomes even more so as the Web and networks evolve as social media, with vastly larger groups of people posting material and creating viable communities online. Accounting for Web n.0 as an ever-escalating series of developments and redevelopments, this chapter offers a series of linked impact reports on the culture machine's electronic environments.

Taxonomics

No matter the name, systems theorists have characterized the emergent Web as displaying robust architectures of participation, having evolved into a truly social software, with a myriad of new ways to link people together. The characteristic usage of the Web in the 1990s was surfing from one static Web page to another. The contemporary Web offers a more dynamic experience in which the users themselves contribute to the environment. Wikis, blogs, and networking sites create affordances for an ever-expanding number of people to share their experiences, perceptions, and productions. Sites like flickr, for tagging and sharing photos, and del.icio.us, for social bookmarking, allow users to categorize, collect, and share their archiving strategies, and has even led to a new term for this explosion of user-generated activity: "folksonomies."[1]

The opposition here is between librarians, archivists, and information specialists, who all professionalize and systematize this kind of activity into "taxonomies," and the evolving personal and social-group-driven folksonomies. The taxonomists are those who have devoted their lives to sophisticated systems for categorizing and organizing the world, drawing on predecessors from eighteenth-century Swedish biologist Carolus Linnaeus and his rankings of class, order, genus, species, and variety, to Melvil Dewey's nineteenth-century decimal system for arranging books on library shelves, still in modified use today.

The folksonomists, by contrast, tend to invent their categories, and even more important, create the tags, terms, or keywords associated with a file or digital object. As these tags are linked through the network, they become markers in ever-larger systems, adding levels of what has come to be termed "metadata." The downside of folksonomies is that they lack clear rules: people tag pictures of their cat with its name or what the animal may be doing, rather than as *felis catus*, the National Center for Biotechnology Information's taxonomy identification number 9685, with a Global Positioning System marker and a time stamp. Of course, even the taxonomists assume multiplicities, with *felis catus* having equally valid synonyms like *felis domesticus* and *felis silvestris catus*.[2] Add to this the fact that there are geometrically more amateur folksonomists than there will ever be rigorous taxonomists, not to mention that the National Center for Biotechnology Information taxonomy number will almost certainly be of less interest to most people than some other, more random, metatag (the cat's cuteness or ability to use a toilet come to mind).[3]

As they layer complexity and even confusion into expanding networks, folksonomies expand affordances for uploading and comprise a net social good. Social media sites like MySpace and Facebook enable users to create giga-, tera-, and even petabytes of new data in the form of texts, pictures, sound files, and video. In social media environments, posting becomes easier, while finding, sorting, and storing become ever-more complex.

The relentless push to market technological innovation helps drive new habits of mind like folksonomies, but also places attractive impediments in their way. Two of the present grails are ubiquity—the embedding of computational power in every environment—and mobility—the ability to communicate with the network from anywhere. Together, fully mobile, ubiquitous computing could make downloading that much easier, but the transition to handheld and telephone devices could make uploading that much harder. The reason for this is that we tend to sacrifice input capacity for size, mobility, and ubiquity. As keyboards, screens, and even lenses get smaller and smaller, discourse tends to revert to the text-messaging level—shorthand like CUL8R for "see you later," acronyms along the lines of IMHO for "in my humble opinion," and the ever more obscure encryptions like CD9 for "Code 9: my parents are watching." There are huge benefits to the culture machine of this ubiquity and downsizing; the capacity to upload photos and video is just the first and most obvious advantage. Yet there are also commercial logics at work here that would turn the device into a permanent distraction and shopping affordance—a technologically augmented "magic shopping wand" that allows us to buy things with a swipe.

If the recent Web, linked to desktops and fully featured laptops, becomes an anomaly on the path to an ever more one-sided consumer mobility with voice telecommunication added, we will have made a major mistake. We should not sacrifice the capacity to upload for the possibility to download.[4]

This caveat is central to the development of not just Web 2.0 but also Web n.0. As computing shrinks, goes mobile, and then envelops us in augmented, ubiquitous environments, users become less and less aware that they are actually "using" something. This is even more of an issue when users are not "using" computers at all but instead are using mobile phones, or enmeshed in invisible networks and activating unknown or even unknowable interfaces simply by being present in a particular place and time. The infrastructure for all of this may well be invisible, yet precisely because it will be hidden away, the questions of what affordances it provides will be far more than technological issues. They will get to the heart of the political battles over downloading and uploading. When the Oxford Internet Institute's Jonathan Zittrain titled his book *The Future of the Internet and How to Stop It,* he was being ironic, but the issues he was concerned with remain very real.[5] The market tends to favor an interactivity restricted to two choices: buy now or buy later; a dedicated salesperson never takes "no" as the final answer, after all.[6] When the mobile, networked, interactive communication device is reduced to a portable television or shopping wand, the war between downloading and uploading will be well and truly lost.

Battles over infrastructure are often unseen by the general public, since they are fought between competing commercial, regulatory, community, and nongovernmental groups all hoping to influence how a society will deploy its resources. Providing water and eliminating waste are infrastructural issues that go back millennia, and the stakeholders can draw from vast historical precedent. When new technologies emerge, the early infrastructural regulation can frequently come from the community of makers and enthusiasts, as it did with ham radio and even pioneer aviation. As time passes, and media and mechanisms move into wider use, though, the early structures for regulating infrastructure break down.

The development of the infrastructure for the networked culture machine involves a huge number of actors including sovereign governments, national and international advisory panels, and affected industries. Even something that sounds like it should be simple, like the assigning of domain names, becomes a political issue, reminding us that struggles about technology may begin technically, but they become social quickly, especially in networks. And it is in harnessing the power of networks and the people who constitute them that Web n.0 is at its strongest.

▾ **SIDEBAR**

The Woman versus the Bulldozer

In 1956, a woman stood down bulldozers sent to flatten Greenwich Village's Washington Square Park. Robert Moses, the great builder and most powerful man in New York for a generation, had decided that the city needed a new expressway to link the lower parts of the east and west sides of Manhattan, and that the patch of green on his map that represented Washington Square Park would be a good place to run a road–making it easier to get from New Jersey to Long Island, and vice versa. Moses being Moses, first he sent in the lawyers, then the city commissioners, and finally the bulldozers. What the master of all twentieth-century master planners had not counted on was one woman who linked her passion for cities with an expressive prose style and a willingness to act. Her adversaries came to regret underestimating her unassuming look. A housing administrator who battled her once said, "What a dear, sweet character she isn't." She went to jail twice for defending her neighborhood, and was able to work with a large group of people who questioned why cars and commuters were more important than parks, communities, and pedestrians. The woman decided to write down the record of her experiences and thoughts about cities and urban planning, and the field of urban planning was changed forever. She was Jane Jacobs, the year was 1961, and her book was *The Death and Life of Great American Cities.*

Jacobs was our preeminent urban anthropologist–a person who could look at a city block, and through building up the details, show exactly how it worked. An associate editor of *Architectural Forum* in the 1950s, she became more and more concerned with the deadening effects of urban planning on cities. She went over the whole sad history of those influential thinkers who saw cities as horrid, dirty, overcrowded places filled with the dregs of humanity who needed planners to come in and rationalize, de-densify, and order their spaces for them. Jacobs instead looked out her window and analyzed what worked in cities, seeking those "fine grained mixtures of street-uses" that enliven any great city. She valued mixed-use areas, where people live, work, shop, and play in contiguous spaces, at discontinuous times. As urban renewal destroyed these kinds of neighborhoods in favor of single-use ones—think of the Lincoln Center arts complex in New York City or the Cabrini Green housing projects in Chicago—cities no longer were able to knit themselves together as well as remain safe, enjoyable, and viable. Jacobs wanted stimulating, mixed-use cityscapes to enhance urban economic actants. These urbanites develop technologies, export them out of the city, and establish cosmopolitan habits.

Jacobs's attention to detailed observation, to bottom-up rather than top-down modeling, and her attacks on monocultures of all kinds commend her work to us. Though jeered at by professional planners of her day—one dismissed her work as "bitter coffee-house ramblings"—Jacobs has certainly had the last laugh, with *The Death and Life of Great American Cities* utterly upending town planning for more than fifty years through its articulation of precisely what makes a neighborhood worth inhabiting. We will spend at least another generation working out how Jacob's fine-grained mixtures should function within digital environments, but mining her work for insights into the culture machine does not stop there. Just after the fall of the Berlin wall and the breakup of the Soviet Union, Jacobs wrote *Systems of Survival: A Dialogue on the Moral Foundations of Commerce and Politics*, in which she identifies two complementary and opposing moral syndromes: one based on taking (also known as the guardian syndrome), and the other based on trading (or the commercial syndrome). These two are sometimes mutually reinforcing and sometimes in grave opposition, but both are required for a viable culture. The guardian syndrome can be the source of stagnation and oppression, but Jacobs also identifies it as the font of art. The commercial syndrome guarantees material progress and undergirds democracy, yet tends to honor nothing that cannot be traded. Without a proper guardian infrastructure, the commercial moral syndrome could be destroyed.

For Jacobs, the worst situation is the creation of the moral hybrid, a commercial group with guardian powers, as when a criminal syndicate like the mafia comes to dominate a society, or guardians with commercial aspirations, like the Chinese Army's control over local industries resulting from the market-economic reforms after Mao Zedong's death. These syndromes are complex agglomerations of attributes, attitudes, and symptoms. In that, they serve as a model for the ways in which we will talk about downloading and uploading. Jacobs was a champion of hybridity, but understood that the secret was to maintain the right balance of the elements and system. Jane Jacobs is inspirational in terms of reminding us that deep systemic analysis can be linked to action.[7]

Metcalfe's Corollary

These infrastructure battles become more and more important because as complex systems evolve over time, what gets constructed now, no matter how ad hoc, tends to be grandfathered in as time goes by. Bob Metcalfe—coinventor of the Ethernet technology, founder of industrial giant 3COM, and a pioneer in wiring people together—put forth one of the most succinct analyses of networks ever offered: the value of

a network is proportional to the square of the number of users of the system (*n*2). Networks become more powerful and valuable as more users join them. One fax machine is useless, but two fax machines create a secured connection, and the more fax machines that are introduced into the network, the greater the value to each individual sender and receiver—a geometric rather than arithmetic increase with each new user.

The cyberpunk maxim that "the street finds its own uses for technology" often enters the discussion at this point. In other words, the intention of the makers is frequently contradicted by the choices of the users, and as more users enter a network, Metcalfe's law indicates that they will be affecting it geometrically. I would propose a corollary of Metcalfe's law that applies to theories of technology as much as the original does to the technologies themselves. Metcalfe's corollary would be that *ideas* about the way technologies operate become that much stronger the more they are sited throughout the user bases of those technologies. The more people come to see what they are doing with computers in terms of a desktop, the more they accept visual icons and habituations of digital-interaction that make reference to physical desks. As they come to see themselves as defined by their connection to the network, the more ubiquitous connection becomes the expected or even "natural" unimodern state.

This means that the way we establish, regulate, and expand our infrastructures affects not only the quantities of goods and media that can flow through networks but the very memes that inform them. The war between downloading and uploading, and the move from Teflon to sticky media culture, can thus be seen as meshworks of intersecting and competing memes.[8] The next order of business, then, is to develop models for the analysis and critique of such systems.

▾ **SIDEBAR**

Mickey Mouse Wants to Live Forever

We traded immortality for sex millennia ago when endlessly self-replicating single-celled organisms mutated into multicellular constructs that began to breed and reproduce new members of the species. Evolutionary biologists theorize that complex, sexual organisms are programmed to die at precisely the point that their offspring no longer need them to survive—the very threshold at which parents would be competing with their young for resources.[9] So is nature bounded, but in culture we are starting to see the return of immortality, at least in the guise of copyright laws. Mickey Mouse, for one, is quite seriously questing for life immortal, and seems to have the means to live his dreams.

No multibillion-dollar enterprise has so identified itself with a single iconic character as has the Walt Disney Company. The mouse ears brand is differentiated into thirteen separate subbrands, but they all relate back to the ur-mouse. Disney debuted the archetypal little rodent in the *Silly Symphonies* cartoons in 1928. From then on, Mickey stood for a playful, lighthearted Americanness that was admired the world over.

From Soviet filmmaker Eisenstein to European theorist Walter Benjamin to Hollywood farceur Preston Sturges, Mickey was central to the cultural imagination, both high and low.[10] In *Sullivan's Travels* (1941), Sturges addresses the existential crisis of his director protagonist: a man who has made a fortune making comedies, but feels spiritually unfulfilled. Sullivan ends his quest in jail, and learns his lesson by watching his fellow cons (a multicultural lot) laugh after a long day on the chain gang at none other than Mickey and his antics.

The problem is that the Disney Company wants to extend its control over Mickey into perpetuity, violating not just the letter of the law, but its intent. By holding on to Mickey–and here the mouse stands for all the characters, songs, stories, and icons that are presently sequestered behind phalanxes of high-paid lawyers–the corporation keeps the rest of us out of the storehouse of mutable materials for the creation of new, noncorporate culture. This is precisely the sort of cultural production that the computer not only makes possible; it is what it all but demands of us as a society.

Creative Commons Culture

There is a primary difference between material goods and intellectual property. A real estate property exists within set boundaries. It can be subdivided, built, up, redeveloped, and so on, but is still very much bounded in space. While it can be developed and redeveloped, real property is neither replicable nor infinitely divisible. As we move into the realm of mass production, the attributes of real property diminish slightly even though their materiality remains. Intellectual property may result in real property, but at base it is still the only form of exchange in which the producer can sell "something" to a consumer and yet still hold on to that something.

Again we find ourselves confronting problems of plentitude rather than scarcity. The cultural archive is overflowing as it never has before. The Royal Library of Alexandria and even the Library of Congress pale in comparison to what has been coming online over the past decades. But as we move further and further into multimediated forms of writing, we have to

make sure that present holders of copyright do not stifle those who would use the networks as a culture machine.

It has been less than a century since entertainment morphed into intellectual property. It was in that century that so many seemingly indelible icons emerged, from Rick's Café in *Casablanca* to Tintin to Hello Kitty to Darth Vader, and, hovering over them all are the ears, gloves, and profitability of Mickey Mouse. If we want to stake out a nuanced position in the discussion over copyright, we can begin from a standpoint that acknowledges that the culture machine exists in a capitalist framework, and that it is foolish to oppose the idea that people should be able to profit from the production of ideas as much as they do from the production of objects. As well, the battles being waged over these issues will not make multinational conglomerates abandon the business of culture (there is simply too much money to be made). In any case, large media companies can promote the production of new, potentially marketable culture by creating incentives, environments, and affordances for fan-generated content. What should be examined and challenged in this situation is the disproportionate power that accrues to large corporate entities when they seek to push copyright into infinity.

In 1998, the U.S. Congress enacted a copyright term extension act, which came to be known as the Mickey Mouse Protection Act. This act added twenty years on to intellectual property protection and seemed to offer even more, ad infinitum. The legislation was widely interpreted as being overgenerous to large corporate interests as opposed to thinking through the general importance of a free flow of discourse through society. Groups like Creative Commons are working against these trends, preferring, in its words, to "reduce barriers to creativity" and build an "intellectual works conservancy."

Creative Commons is an initiative that was founded to help "people dedicate their creative works to the public domain—or retain their copyright while licensing them as free for certain uses, on certain conditions."[11] Rather than software, Creative Commons focuses on writing, music, and art to be developed and distributed under different intellectual property models than those exercised under the current copyright laws. Leading lights in the Creative Commons movement, such as legal scholar and activist Lawrence Lessig, feel it is imperative to address the arrogance of companies like Disney and News Corporation's Fox and Time Warner, which use their economic and political power to extend their own copyright seemingly into infinity.[12]

At some point an idea, situation, or character becomes prevalent enough to form a meme. Society then benefits from its open use by all as opposed to restrictive ownership by the private producer. The open-source cultural movement has many diverse gradients, from those on the far "copyleft" side who oppose any form of protection or privatism whatsoever, to those who want a blend of protection and the eventual diffusion of knowledge. This kind of argument does not interest the copyright lawyers hired by Time Warner, Disney, and all the rest, but they are narrow stakeholders in an evolving landscape. The original intent of copyright laws was not solely to reward the holders of these rights but rather first and foremost to encourage innovation. The transformation of intellectual production into intellectual property through the course of the twentieth century skewed this, yet we must remember that the entire purpose of copyright was to encourage uploading. We need to reboot this as an assertion. Abusive intellectual property rights lock down culture.

The cultural as opposed to technical implications of the open-source movement began to be felt through the 1990s, as ever more people wired themselves to the Internet. Artists and in particular net.artists displayed a natural affinity to the open-source movement. Since that time, there has been an explosion of peer-to-peer file-sharing networks (from Napster to Morpheus to BitTorrent to the Pirate Bay) and a backlash from the "content industries" (including an infamous suit brought by record moguls against a twelve-year-old girl). But regardless of the ever more divisive debates in the courts, the reality is that open-source cultural production is in the ascendance. Take the case of Jenny Everywhere aka the Shifter. Jenny is an open-source comic character who can be developed by any artist who wants to use her. According to the Web site where she was born and lives her (many) lives: "The character of Jenny Everywhere is available for use by anyone, with only one condition. This paragraph must be included in any publication involving Jenny Everywhere, in order that others may use this property as they wish. All rights reversed."[13] The open-endedness, the unfinish of Jenny Everywhere, distinguishes it from the similar neologism "crowdsourcing." Crowdsourcing is also about the deployment of multiple, online "eyeballs," but the concept's link to the "outsourcing" of globalization ties it tightly to the economic realm.[14] Imagination, on the other hand, encompasses but is not limited to those projects that can be monetized, so it is less "problem solving" than "situation enabling" that is needed.

The Creative Commons movement is trying to ensure that those who want to "write" with the culture machine—by this, I mean make and distribute motion pieces, new music, filmic fictions, digitally modeled fabrications, ubiquitous information environments, photo blogs, and the list goes on—will be able to have access to the contemporary raw materials of creativity. Whether you choose to call this remix culture, as Lessig does, or simply the new way of writing, what the Creative Commons, free culture, and *culture libre* movements and thinkers are ensuring is that just as authors in the past had the right to think about earlier forms and writings, and incorporate, comment on, and build on them—not just James Joyce's references to Homer, but the incorporation of actual newspaper fragments in Georges Braque and Pablo Picasso's seminal collages of the 1920s—so today's artists, writers, filmmakers, bloggers, musicians, and designers must be allowed similar access to the material that informs their creativity.

The difference between the 1920s and today, however, is that while the ease of copying has indeed posed new problems, the strengths of copyright holders has grown enormously. The ubiquity of the search engine has not only made a plethora of material available to potential users but also has made it concurrently easier for the owners of intellectual property to track down copyright "violations." In their zeal to protect their intellectual property, corporate forces have consolidated control over political processes. The result is that government regulators respond more quickly to the concerns of those who already have money much faster than they do to the forces that become empowered by technological innovation.

The innovations of the culture machine are bringing about changes in technological systems, the content transmitted by those systems, and the aesthetic quality of that content. These innovations will require an evolution of the legal system to serve as a counterbalance to the consolidated intellectual resources of established hierarchies, centralized governments, and transnational corporations.

▾ SIDEBAR

© *Ulysses* -vs.- 🅭 *Odyssey*

In *Ulysses*, James Joyce has his protagonist/alter ego Stephen Dedalus proclaim that a "man of genius makes no mistakes; his errors are volitional and are the portals of discovery."[15] But what of the sons and grandsons of genius? Are their errors equally productive? Looking at the legal battles over copyright that Stephen James Joyce has waged in his ancestor's name, the answer appears to be no. The case of the Joyce estate is a cautionary one. The modernist, like many great artists, was ambivalent about literary analysis and history, but he explicitly constructed his two major novels, *Ulysses* and *Finnegans Wake*, in ways that would engage critics for decades after their publication. Joyce judiciously engaged with those who came to embrace his aesthetic agenda. These critics and professors in turn made Joyce's work the sine qua non of the Western literary tradition, guaranteeing that students would wrestle for decades to come with Dedalus, Molly Bloom, and Finnegan's corpse. This dialogue between author, work, critics, and students was hardly open, but it was vibrant and did generate a significant revenue stream for Joyce's descendants.

Paradoxically, many of the great moderns became defining artists and writers precisely because they were among the most gifted "remixers" of their era. We have already seen in chapter 3 how this worked for Duchamp and his ready-mades, but the same is true of Joyce, who drew from the whole of the literary canon, and a range of languages and recombinatory puns, especially in *Finnegans Wake*. Joyce took the "plot" for *Ulysses*, that famous novel of Ireland, from that famous epic of Greece, the *Odyssey*. If Homer's heirs had inherited the same legal structure as Stephen James Joyce, would they have sued?

By the start of the twenty-first century, however, the most powerful overseer of the estate of the twentieth century's greatest literary modernist was his grandson. Stephen James Joyce's restrictions on the critical, historical, and general use of his grandfather's work have brought "Joycean" activities to a virtual standstill. Stephen James Joyce objects to biographical scholarship into the extended Joyce family, so he has closed off the family archives. He has used the riches of the estate to sue and thereby silence scholars and journalists with whom he disagrees. He has even denied the city of Dublin rights to read selections from *Ulysses* during Bloomsday celebrations. The overall effect of this is not just chilling but also can be seen to be economically disastrous. New scholars are actively discouraged by their mentors from working on Joyce because there is nothing that they can do about the estate's prohibitions. Without new scholarship and even general criticism, Joyce slowly fades off syllabi, and Molly's famous soliloquy, "yes I said yes I will Yes," is choked off by the "no I said no I will Not" of overzealous copyright defense.

CHAPTER FIVE

BESPOKE FUTURES

▶ notes: pp. 190–194

Bespoke Futures

The secret war between downloading and uploading is not being waged in isolation. It must be contextualized within a much wider and less metaphoric series of battles worldwide, stretching back decades. A key feature of these conflicts has been the oscillation between irrational exuberance and untethered terror. To delineate the period we are living through, I combine the hopefulness that followed the fall of the Berlin Wall with the fear and rage that followed the attacks on the World Trade Center in 2001. This period I call 89/11 (pronounced "eighty-nine eleven"), and this section of the book will both define that era's characteristics and move past the stasis it engendered via the creation of what I term "bespoke futures."[1]

What looked like it would be a facile history in 1989—the victory of one sort of built system over another, the triumph of democracy over totalitarianism, capitalism over a command economy—turned out to be vastly more complex. The post-1989 period contained a multitude of features, but one unifying construct was the belief that after the fall of the Berlin Wall and then the Soviet Union itself, not just Communism, but all the countervailing forces against market capitalism were vanquished, and not just for the moment but literally for all time. The Market with a capital M was the grail at the end of Francis Fukayama's treatise *The End of History.*[2] The Market was the solution for all questions, the Market would bring peace and prosperity, and would free itself from the tyranny of the business cycle, evolving into an entirely invisible, frictionless, perpetual motion machine that would take the name of the New Economy (again with capital letters).[3]

This immediate post-1989 period coincided with the most utopian phase of the culture machine: the euphoria of the World Wide Web's first Wild, Wild West phase. For close to a decade, people talked about bright, posteconomic futures, perfectly transparent global markets, and the glories of pure, unadulterated information flowing around the world at the speed of light. In the view of those who came to be characterized as neoliberals, digitized networks connected the far-flung commercial enterprises of the lightly regulated and purportedly stateless world. Information was simultaneously weightless and powerful, with the premillennial moment promising limitless economic and social possibilities, at least to those wired into the network.

Even beyond the rarefied worlds of venture capitalists, newly minted high-tech millionaires, and those who "cashed out" on their stock options (now rich enough to consider themselves posteconomic), there was a generalized optimism about the market effects of computers. Much was made of productivity gains, and economists pointed to the maturation of information technology, within the economy, from Walmart's sophisticated stocking softwares to small businesses' use of efficient databases.

The rhetoric of freedom and empowerment ranged from cyberlibertarian individualism to communitarian networks, but the first wave of enthusiasm for the World Wide Web did promise open, common, and often free culture at the same time that the unimodern communication technologies spread the memes of advanced capitalism and market liberalization around the globe. Certainly there were the occasional glitches in the system, as when antiglobalization protesters disrupted world trade gatherings in places like Milan and Seattle, but for the most part the trend lines (as business forecasters like to call them) soared upward into an ever-bluer sky. With Communism dispatched, no other ideology could challenge the market—a market that now, through the addition of the simulation and participation modes of the culture machine, was claimed to be impervious to the dislocations of earlier business cycles. But this period was shorter lived than even the Seattle protesters expected it to be.

The NASDAQ, a U.S. market heavy on high-technology firms, was the most important index for the New Economy. It crested in March 2000, and within a year had lost more than half its value, vaporizing trillions in paper profits. The stock market losses for three companies alone—AOL, Amazon, and Yahoo!—amounted to three hundred billion dollars.[4] The indexes continued to sink for the next year, and then came the critical event that signaled the complete end of the period inaugurated with the dismantling of the Berlin Wall in 1989. It was on September 11, 2001, that the markets took their next massive hit, and the newness of the so-called New Economy had its last bits of hype sucked out of it.

For Americans, at least, the faith in the Market to overcome all obstacles suffered a near-fatal blow that day, not simply with the vivid reminder that history in fact had not ended, but also that all those high-flying young engineers, venture capitalists, and entrepreneurs in their Casual-Friday-Every-Day chinos and polo shirts were now being edged out of the spotlight to prepare for the return of the Blue-Suited-Wingtip-Shod-Flagpin-Lapelled grown-ups (think former Vice President Dick Cheney and Secretary of Defense Donald Rumsfeld).

These were cyclic booms and busts, of course, and the advent of Web 2.0 and its implications of Webs n.0 in the future have reignited a lesser technofabulism. No one talks of social networking sites as being "more important than fire," as they did with Web 1.0.[5] That is because fire, or at least FIRE, came roaring back in the even bigger financial crisis that followed the bursting of the Web 1.0 bubble. FIRE—an acronym for Finance, Insurance, and Real Estate—was enabled by the increasingly frictionless global networks of information, commerce, and trade in the first decade of the twenty-first century. The FIRE financial crisis was less concerned with the mystical powers of the technologized market than a retreat into old-fashioned Greed with a capital G fueled by debt (and the dissolution of oversight and regulation), though the innovations in fiscal instruments that sustained the aura of FIRE's invincibility relied ever more heavily on computers to crunch numbers and parcel out risk.

The FIRE fire sale only intensified the loss of faith in the Market with a capital M, but this is not to say that it has been replaced with some alternate economic or social certainty. Before the events of September 11, 2001, I wrote that "with tribalism and fundamentalism appearing to be the only other options on the political scene attracting adherents, postindustrial capitalism would seem at this point as inevitable and all powerful to the artists of the West as the Christian Church must have been to artisans of 11th century France."[6] The attacks on New York, Washington, Bali, Madrid, and London along with the reverberations thereafter in Afghanistan, Iraq, and around the world only prove this point. For those coming of age in the 89/11 world who choose not to embrace the rigid certainties of tribe or faith, a viable alternative to capitalism has simply not manifested itself, no matter how fervent the desires of antiglobalists manifested on the street. The 89/11 theories of aesthetic production, then, begin with the market and its forces, but must not end there.

It would be pleasant to retreat into the irrational exuberance of the technofabulist boom and see within it the triumph of the digital culture machine. But as noted, an H-bomb smuggled on a container ship or a human bomb walking through a crowded cafe resuscitate the dread of the annihilation that characterized the period from the atomic bombing of Hiroshima through the end of the cold war. The 9/11 mentality, unfortunately, internalized this terror, reworking it into a echo chamber of fear and isolationism, and one that for too long dominated the tenor of television's twenty-four-hour news cycle and then accelerated even faster as the blogosphere gave birth to nanocycles of information. To make a positive contribution, moving past the 89/11 manic-depressive swings, means replacing fear as the default setting, without succumbing to nostalgia for a market triumphalism that cannot sustain itself.

How can the computer as culture machine save us from fanatics of all stripes and faiths? From the smart chips in toasters and washing machines, to the Global Positioning Systems in cell phones and cars, to the invisible webs of closed-circuit television surveillance systems, we are ever more enmeshed in computational and communicative networks. As all of these objects and environments become inexorably smarter as well as interconnected, what we mean by culture itself expands. We are not just talking about video blogs and fabbed knickknacks; in a lived environment where unimodern informationalism has taken hold, the realms of nature and culture fuse, insides are connected to outsides, and culture machines envelop us completely. This is what gives the computer the chance to match and even overpower the bomb.

The science-fictional author and futurist Bruce Sterling believes that neither utopia nor oblivion lie in our futures, but instead blends them into the neologisms "ublopia" and "otivion."[7] His point is that there are places on the planet where the apocalypse is already happening, and other places where the perfected future already seems so firmly entrenched that it is impossible to imagine it as otherwise. The computer will be central to much that is awful now and in times to come, but contains within it the way out of the very problems it ignores now and exacerbates, in large measure because the culture machine allows us to visualize and understand our lives not simply as a series of events but instead as a system of systems. One word to describe building this "system of systems" is "design." Designer Bruce Mau, who coined the phrase Postscript World noted earlier, did a famous napkin sketch that features a figure/ground toggle between design and society. He draws two images of concentric circles next to each other. The one on the left has as its outermost ring the word "nature," next in is "culture," then "business," and finally nested at the center is "design." This sketch shows design in its historical role as the servant of commerce, as reflected in the early term for graphic design, "commercial art." The right-hand circle shows what has happened, in Mau's view, since the advent of nanotechnologies, genetic manipulation, large-scale digital fabrication, and even terraforming. Design expands to become the outermost ring, encompassing nature, culture, and business. From Mau's perspective, humankind's increasing capacities to manipulate and shape everything from the submolecular to planetary scales radically transforms what design means and signifies—in his words, that design itself "has become the biggest project of all."[8] Mieke Gerritzen, a Dutch visual provocateur, went so far as to proclaim that "Everyone Is a Designer!" in a book she designed by the same name.[9] Whether we accept the hyperbole of her manifesto or not, there is much to be gained from at least thinking like a designer in this 89/11 world. Designer Michael Beirut equivocates perfectly: "Not everything is design. But design is about everything."[10]

▾ SIDEBAR

Sears -vs.- VDNX

In 1939, in a six-hundred-acre park in the north of Moscow, the Exhibition of the Achievement of the Soviet People's

Economy opened. Known as VDNX, this propaganda park was a Potemkin village, a trade and technology fair, and a model farm all wrapped up into one. VDNX was a phantasmagoric space where the Soviet iconography of happy, healthy workers, powerful tractors, and glistening satellites was mirrored by the bounty of prize pigs and luscious produce. It was a central showplace of the Stalinist spectacle, and truly fit Maxim Gorky's idea of socialist realism as revolutionary romanticism.[11] VDNX showed life not as it was lived but rather as it ought to be lived.

That same year, thousands of miles from Moscow, a Sears store opened in Los Angeles at the intersection of Pico and San Vicente boulevards. This was one of the retail giant's crown jewels, a department store where the building was constructed from the ground up to showcase the merchandise. This syncretic construction was a fairly novel concept in which the tables, fixtures, space requirements for the different merchandise lines, customer flow and width of the aisles as well as the building's shell itself were all built around the selling floor plan. At the time, a rival merchandising executive offered tribute: "In my long experience in the retail field," he said, "I have yet to witness a . . . unit which equals Sears Pico Store in practical efficiency, merchandise engineering, operation, layout and presentation of merchandise."

Setting the story up this way appears to augur a classic dialectic—one pitting Communist exhibition against capitalist showcase. After the events of 1989, the conclusion would appear obvious: the victory of the market against the failures of the control economy. This would mirror the self-congratulatory prose emanating from the pages of the U.S. media. The tolerant, antistatist, neoliberal tone of established media like the *New York Times* cried out for lampoons, though none were forthcoming in 1999, a year in which triumphalism dominated (in large measure because the New Economy was so dependent on the fantasies of political and technological omnipotence). The histories we inherit tend to be the stories of conflicts as written by the victors.

Discursive excesses aside, 1989 was of central importance to the way we make culture and think about history decades later. That year saw the Czech Velvet Revolution, the fall of the Berlin Wall, the reunification of Germany, the eventual fissioning of the Soviet Union, the emergence of the Baltic states, and the continued extension of market reforms in China (which coincided with the political repression of Tiananmen Square). The Polish trade unionist, journalist, and now capitalist newspaper owner Adam Michnik put it well when he noted in a commemoration that "the revolution of 1989 was a great change without a great utopia."

So let us return to VDNX and Sears today, after this "great change." By the mid-1990s, the USSR was no more, and north of Moscow, in a city once again in a country called Russia, VDNX was transformed through that peculiarly post-Soviet mix of perestroika, privatism, and gangster capitalism. As one observer commented soon after the dissolution of the Soviet Union, the "exhibitions pavilions, built as palaces for the people, have been transformed into communal apartments of commerce: VDNX is now a bizarre shopping mall. Many of the most opulent pavilions have become congested labyrinths of tiny stalls that sell a jumble of consumer goods." By the turn of the millennium, the Space Exploration Pavilion was full of used cars, although there were a few satellite and rocket models hanging from the roof above them. The less grandiose pavilions had been rented out by new Russian companies, many of them protected by private guard services against Russia's rampant gangsterism.[12]

But if this is to be the story of triumph, we must follow VDNX's rival in Los Angeles. The only problem was that by the mid-1990s, the building had been sold and was no longer a Sears. Now it stands where the Central American, Korean American, and African American communities meld in downtown, the Crenshaw district, and the southeasternmost edges of the rich Westside. The first floor was taken over by a massive but always-understocked discount hardware store. The second floor became the Pico Swap Mart. Most of the carefully designed walls were knocked down, and the whole space was cut up into a series of cubicles that were separated from each other by metal fencing. Small shopkeepers, primarily Koreans and Guatemalans, filled each little space with a profusion of the sort of off-brand, odd-style goods that you expect to see in Lima, Manila, or Marrakech, but not in the heart of the standardized, homogenized United States. In fact, there were probably similar off-brand items available at both VDNX and the Pico Swap Mart. As for the third floor of the former Sears, it was simply closed off by more of the chain-link fencing. Since that time, it has been torn down and rebuilt as a mall anchored by a home improvement superstore, awaiting the next stage of capitalism's creative destruction.

Mutants and Modernists

Close to two decades of teaching and interacting with artists and designers has prompted me to think that design should now be seen as an expanded discipline. As a bonus, design as meta- or even megadesign can help to address the vision deficit that the 89/11 period has saddled us with. I realized how serious this deficit was one afternoon when an extremely

talented student asked me to look at his sketches for a film project. The assignment was to create a scenario and art direct a science-fictional film about the future. There were no other requirements. He was free to imagine as widely and wildly as he wished. This is what he came up with:

> My film idea is set in Southern California. Radiation from the nuclear fallout has mutated most of the surviving humans, and those few who have proven immune . . . made the remains of the former Rose Bowl their home. To protect themselves from the violent and mindless ex-humans, they have fortified the Rose Bowl's walls and made the interior . . . completely self-sustainable.

The architectural illustrations were exceptionally detailed, and the mutants appropriately deranged looking, but in the end I was profoundly depressed by the fact that this kind of post-apocalyptic landscape is now the default when we ask creative people to speculate about the future. I started to think about why we have so little faith in the future. One reason is that the shape of things to come has never been so inadequately imagined.[13] Knock modernism, if you choose, but at least the art, design, and architecture generated in that heady era put forth a panoply of futures seductive enough to inspire others to bring them into being. The twentieth century offered a surplus of futurities, or those qualities we associate in or with the future itself. In the twenty-first century, though, we seem to suffer from a vision deficit, an inability to imagine a future or futures that we would actually like to live in. What is needed is something to quicken the heart about the future, something to invest us with hope, excitement, vision, and will. In other words, where are our jet packs?

This lack of vision about the twenty-first century is in direct contrast to the explosion of imaginings about the twentieth.

From the turn of the nineteenth century through even the first decades of the twentieth, the last century was explored and envisioned via world fairs, expositions, Sunday supplements, and popular fiction. In 1968, director Stanley Kubrick made *2001* and created one of the signature attempts at envisioning our century. Even though he made the film at the tail end of the 1960s' expansive, postscarcity, hippie-influenced, space race fever dreams, Kubrick's vision is already melancholic, portraying the coming century as even more banal in some ways than his present. Only the alien technology reinvigorates wonder. A decade and a half later, Ridley Scott eliminates even the cosmos as a source of inspiration, with the twenty-first century's "Offworld Colonies" in *Blade Runner* (1984) unseen, except in advertisements floating by on blimps. The success of Scott's "retro-deco" style in *Blade Runner* essentially stopped the popular visioning of the future in its tracks, locking successive popular media futures into a permanent present of ever more encrusted layering of technologies and styles. Cohesive visions of the twenty-first century as a social or even technologically inspiring whole have been rare to nonexistent.

There are good reasons for this vision deficit, of course. The twentieth century, so gaudily imagined in the decades before World War I, confronted the bloody futility of trench warfare in Verdun, followed in turn by the Great Depression, the atrocities of the Nazi Holocaust, the Soviet gulags, Chinese central planning famines, and the boundless ferocity of Cambodia's reeducation camps after Year Zero. Philosopher Karl Popper's hard-won insight into the twentieth century's charnel houses was that "the attempt to make heaven on earth invariably produces hell."[14] One of the ways that the world was able to reinvigorate itself was by shifting its focus to the actual heavens. The quest for outer space was enough to reignite the imagination after World War II, but once that goal was achieved, the exhaustion described above settled over us again.

Besides the bomb's presence on the global scene, there is also another factor to weigh in. The future itself had so fully arrived that it was going to take decades to sort it all out. Captain Nemo's submarine in Jules Verne's *2000 Leagues beneath the Sea* became the scourge of commercial shipping during World War II. The socialist utopian Edward Bellamy predicted television in his best-selling novel of 1888, *Looking Backward*, under- rather than overestimating its significance. From the pulpy imaginations of comics worldwide, the twentieth century saw the realizations of fantasies from British boy-hero Dan Dare's rocket ships, to daily newspaper strip Dick Tracy's two-way wrist communicators, to the sonic booms over Tokyo of Kazumasa Hirai and Jiro Kuwata's pioneering cyborg manga, *8 Man*. Even the relatively sober prognostications of engineers discussed in the "Generations" narrative came true in the most widespread ways. By 2000, neither Vannevar Bush's proto-hypertextual Memex proposal of 1945 nor J.C.R. Licklider's more amusingly named Intergalactic Computer Network in 1963 seemed futuristic. In fact, they defined the presentness of desktops worldwide. It may be that we will never catch our collective breath, but that does not mean that the yearning for more comprehensive visions of the future has completely lost its value.

▾ SIDEBAR

Where Are Our Jet Packs?

In August 1928, the kids who wandered down to the local drugstore or newsstand caught their first glimpse of the new *Amazing Stories*. They were used to seeing extraterrestial monsters, cruising starships, or steely-eyed heroes battling even steelier robots. What did those pulp-addicted readers think of that month's cover, though, so different in tone and attitude?

The illustrator Frank R. Paul had painted a bucolic landscape with a man floating serenely above it, grasping a glowing wand, and wearing a flying harness and pilot's helmet. In a book on the birth of comics, Gerard Jones talks about this image's impact:

> In the archaeology of popular culture, [Paul's cover] appears again and again as a pivotal memory of a generation of moviemakers, science fiction writers, cartoonists, astronomers, futurists, and rocket engineers. . . . In contrast to the monster-filled labs and devastated planets of most Amazing covers, this is a world of sun and security, defined by architecture, science and supremely economical illustration. And in contrast to the terror and grimacing of every face on the covers of the pulps, this man is smiling. In newsstands filled with dread, here suddenly was joy, a safe but unbounded future. Here in the hearts of children who saw that cover was a soft, exhaustless lift into the open, golden sky.[15]

Paul was commissioned to create this image by Hugo Gernsback, one of the twentieth century's most intriguing cultural entrepreneurs, with a remarkable nose for talent. Gernsback was the most important publisher of science fiction in its golden age, and in his pulps the genre emerged almost whole cloth. The covers of *Amazing Stories* and the other pulps were illustration- rather than photo-based, and the role of illustration in opening up the twentieth-century imagination to the future cannot be underestimated. Drawing offers a freedom that the photographic image does not. From the nineteenth century onward, illustrators would be tasked with creating visions of the future, some tethered to the real while others reaching the flightiest of fancies. These evocations of new technologies, communications, transportations, and habitations prepared the way for the made realities of modernism and twentieth-century material culture. They were often juvenile and sometimes crude, but the pulp modernism that Paul and his cohort of illustrators, comic book artists, filmmakers, and special effects technicians created had an impact that outlasted their makers' renown.

In the 1980s, the cyberpunk pioneer William Gibson published a short story called "The Gernsback Continuum," where the narrator mentions Paul's work as the world of *Amazing Stories* bleeds through into the "real world." The narrator looks out into the desert, and sees "zeppelin docks and mad neon spires . . . in gleaming ziggurat steps that climbed to a central golden temple tower. . . . Roads of crystal soared between the spires, crossed and recrossed by smooth silver shapes like beads of running mercury. The air was thick with ships: giant wing-liners, little darting silver things (sometimes one of the quicksilver shapes from the sky bridges rose gracefully into the air and flew up to join the dance), mile-long blimps, hovering dragonfly things that were gyrocopters."[16] Illustration has died off as the font of imagination in the early twenty-first century, save perhaps for those preliminary sketches of 3-D computer graphic animation. The new source is photo-realistic computer graphic special effects. It remains to be seen if a new generation of Frank Pauls will harness these tools in the same way that their predecessors in the pulps did a century ago.

Adopting the Future as a Client

There are risks to thinking about the future through the lens of design. Such activities can bring us dangerously close to technocratic fantasies of rational utopias, allowing us to elide the problems of contemporary life and even squander our energies in the sheer unlikeliness of it all. Not thinking about the future—both seriously and playfully—can be an even worse hubris. One

inspiration is the practice of design itself, where practitioners regularly take on "pro bono" work. This is the short version of the Latin phrase *pro bono publica*, which translates as "for the public good," and while it is often associated with lawyers, designers also regularly take on lower- or nonpaying clientele as a public service.

I have in the past challenged designers to choose the future itself as one of their pro bono clients, and more than that, insisted they consciously choose a better future to be that client.[17] This choice addressed a key complaint of designers: to whit, that they do not get to work on content that is as compelling as their control over form might warrant. Taking on the future as a client allows designers to dedicate themselves to the development of a new kind of humanism. Designers have numerous clients, and cannot necessarily choose them all. But one can choose at least some of them. In any case, the best designers know that the choice of who *not* to work with is as often more important than who one does choose (or is forced to accept, simply to pay the rent). In a moment when sustainability is gaining more and more traction in design discourse, this future-as-client model moves designers past the defaults of nontoxic inks, recyclable consumables, and walkable cities into deep issues of sustainability, or the very future of design as a human activity.

This is all well and good, but how to adopt the future as a client, what methods are available, and how can these methods function beyond the scope of traditional design and interest those of us who are not designers? One methodology worth exploring is scenario planning, or as will be explained later, the crafting of bespoke futures.[18] Remember that not only have corporations not forgotten the future, they have developed measures to plan for it. Over the last quarter century, farsighted multinationals have institutionalized scenario planning to ponder upcoming conditions and their effects on long-term profit and loss. By taking this corporate scenario planning and perverting its methodologies, audiences, and outcomes, we can transform vision deficits into surplus futurity, proactively filling the present's imaginative gaps with an abundance of wonders to come. Not only is this necessary at our particular moment in time, this bespoke futures process is enabled, as we see in this chapter's section on what I call massively synchronous applications of the imagination (MaSAI), by the advent of visual computing and massively scaled networks. These systems have made it possible for groups of individuals to come together and envision futures they might actually want to inhabit. Indeed, architects and technologists, long the lords of futurity, need to be shaken out of their complacent leadership. The future is about more than just gizmos and skylines (although that certainly still defines our lumpen visions of it); it is about the modes and meanings that people create in these virtual as well as built environments.

In 1985, Royal Dutch Shell's director of planning, Pierre Wack, published a seminal essay in the *Harvard Business Review* titled "The Gentle Art of Reperceiving." Here, Wack outlined the ways in which Royal Dutch Shell—a British-based multinational company—had developed methods for escaping from consensus mind-sets about the future that inform and infect every major corporation. The consensus mind-set is that official future handed down by bosses to subordinates and ingrained within the organization's culture. It has a remarkable staying power, even if subordinates and even a select few in upper management know that following the official future will lead to eventual ruin. The generational slide of the big-three U.S. automobile companies after the oil crisis of 1973 is proof enough that minds, once set, are difficult to change. But it was precisely change that Wack and others were after when they developed a methodology for engaging groups of invested parties in dialogues about the future. They identified change agents, plotted them out on grids, gave potential futures catchy names, and tried not to let the received wisdom of the corporate culture prevent them from thinking the unthinkable.[19]

Scenario planning often identifies five key driving forces at the outset like society, technology, economics, politics, and the environment. The identified scenario drivers are then allayed in a spectrum (along one axis), a matrix (with two axes and four 2-D spaces), or a volume (with three axes and eight 3-D spaces). Some of the better-known successes of the scenario planning process were Royal Dutch Shell's ability to plan successfully for the expansions and contractions of global oil demands after the price shocks of the 1970s, the apartheid government of South Africa developing the capacity to imagine a peaceful turnover of power to Nelson Mandela and the African National Congress, and somewhat less globally significant, the identification and development of a U.S. "gardening lifestyle" by the retailer Smith & Hawken.[20]

Crafting Bespoke Futures

Peter Schwartz and Jay Ogilvy, cofounders of the Global Business Network (or GBN as it is better known), are two of the better-known scenario planners. They have invested a great deal in condensing and distributing the memes of scenario planning. They distilled their experience into "Ten Tips for Successful Scenarios," which are honed from years of working with a huge range of clients.[21] Many of the companies that are engaged in scenario planning are those for whom it is unfortunately already too late. These corporations are thus functioning on the edge of obsolescence, hysteria, or bankruptcy, buffeted by forces and futures they do not understand. The tips that Schwartz and Ogilvy offer are ideally suited for groups with definite goals (profits being the most obvious), with large-scale hierarchical organizations, where decision makers are often far removed from those who actually recognize the change agents on the ground. Hence, much of what they suggest is a script for an encounter group between unequals, in which the less powerful but more knowledgeable and/or sensitive help to draft a compact about the future with those who control their destinies, but not the future of those outside the organization. That caveat offered, here is Schwartz and Ogilvy's top-ten list:

1. Stay focused
2. Keep it simple
3. Keep it interactive
4. Plan to plan and allow enough time
5. Don't settle for simple high, medium, and low plots
6. Avoid probabilities or "most likely" plots
7. Avoid drafting too many scenarios
8. Invent catchy names for the scenarios
9. Make the decision makers own the scenarios
10. Budget sufficient resources for communicating the scenarios

As noted, scenario planners generally work for groups—usually corporations, and occasionally government agencies, nongovernmental organizations, foundations, community groups, or other nonprofits. But what about the idea of scenario planning for the rest of us? In any case, scenario planning, like science fiction, is frequently less about the future than it is about the present—the present's blinkered perspectives and wistful hopes writ large. The idea is to get beyond profit and loss statements, thereby creating an opportunity space for the imagination, and enabling individuals and independent groups to create visions of the future that inspire them. To be clear here, I want to pervert the process, misusing scenario planning to skew toward a goal, a future that I and hopefully others would actually want to work to build. The point is to create a plot that moves from profit and loss to vision and futurity, P&L

to V&F; from a return on investment to a return on vision, ROI to ROV. Like Schwartz, my point is to inject the joy of making and engaged invention into the process.[22]

I propose recasting the whole process, stealing a word from haberdashery, and putting it through a methodological blender. As a British audience is far more likely to know, the term bespoke refers to clothing that is custom made. It comes from the seventeenth century, when tailors held their own stocks of cloth. A customer would come in and choose the fabric for his suit, after which the tailor would mark off the requisite length of material, referring to it as having "been spoken for."[23] So how are we to craft these custom-made visions? Here are ten untested tips, skewed, adapted, and modified from the GBN model in order to create bespoke futures:

1. ~~Stay focused~~
   Stay visionary

2. ~~Keep it simple~~
   Keep it complex

3. ~~Keep it interactive~~
   Design it interactive

4. ~~Plan to plan and allow enough time~~
   Plan for serendipity and allow enough space

5. ~~Don't settle for simple high, medium, and low plots~~
   Aim high

6. ~~Avoid probabilities or "most likely" plots~~
   Fixate on just one scenario that you want to achieve

7. ~~Avoid drafting too many scenarios~~
   Draft enough scenarios to kill all but the best

8. ~~Invent catchy names for the scenarios~~
   Invent catchy visuals for the scenarios

9. ~~Make the decision makers own the scenarios~~
   Own your own scenarios

10. ~~Budget sufficient resources for communicating the scenarios~~
    Generate sufficient fervor to communicate the scenarios

Creatively misusing scenario planning as a means toward crafting visions of the future—often interactive, immersive, or augmented—can inspire us to go back to our own communities or dig deeper into our own creative practices to transform the vision deficit into a surplus futurity. Corporations and governments harness their control over scale—capital and power—to generate their scenarios. A few years back, both Vodaphone and Motorola released interactive, Web-hosted, media design scenarios about the near future.[24] Not unexpectedly, these sites offered an exceedingly technicist futurity—well made, nicely designed, but driven by new stuff, not new ways of making new meaning. These "connection is everything" models heavily promoted by phone and wireless companies are retro-McLuhan: in these corporate futures, the medium always dominates the message. Bespoke futures might well restore some balance. The culture machine offers a growing capacity to create complex visualizations with digital systems and distribute them widely over high-speed networks, engaging with open-source cultural initiatives. We are now capable of taking advantage of peer-to-peer networking, file sharing, and massively scaled distributed computing to develop countervailing forces, or people's rather than corporate scenario-building strategies.

Bespoke Futures as Strange Attractors

At an earlier cultural moment, we might have looked to the, or at least an, avant-garde for surplus futurity and to generate our bespoke futures. But given the use and abuse of the term avant-garde (I have described it as a horse, ridden too hard for too long and in need of an extended cooling-off period), it is important to develop other metaphors.[25] Compare two imagescapes—one still, and the other dynamic. The first is perhaps the most famous diagrammatic representation of the avant-garde: Alfred Barr's 1936 chart, "The Development of Abstract Art." Barr, the founding director of the Museum of Modern Art (MOMA), created a visualization that offers a rational modernist taxonomy of opposition, critical distance, and historical progress to explain and support his curatorial choices at MOMA. If you connect the links one way, you will track the following movement:

Cubism ⟹ Suprematism ⟹ Constructivism ⟹ Bauhaus

Follow another line, and you will get to the Bauhaus this way:

Synthetism ⟹ Fauvism ⟹ Expressionism ⟹ Bauhaus

Barr's chart is a teleological document that culminates with the presentation of these objects in MOMA's galleries. The exactitude of Barr's chart is unlikely to emerge from the process of bespoke futures. The skewing of the classic scenario-building process undermines such vectoral surety and fixed relationships.

Instead, a better model might be found in the dynamic imagescapes of the Lorenz strange attractor, one of the earliest and still most potent visualizations of chaotic systems.[26] Edward Lorenz, a mathematician and meteorologist at MIT, needed a new way to analyze atmospheric conditions. He came up with a dynamic model in which seemingly random and chaotic outliers were eventually contained within a definite figure (often described as looking like an owl's eyes) in which solutions approach but do not replicate each other exactly. The equations are described as deterministic, yet they are extremely sensitive to their initial conditions. This means that it is impossible to predict any single solution at any extended period of time. Pendulums or pistons have relatively simple attractors. More complex systems (like weather, the stock market, or human culture) rely on a huge number of attractors and can be better thought of as "phase spaces." In phase spaces, repetitions and differences lead to constantly shifting equilibriums. A minor change in the original condition can effect a hugely different outcome—better known as the "butterfly effect"—and can also create a different attractor, collapsing it into a fixed solution or tumbling it back into apparent chaos before a new strange attractor establishes itself. This effect is readily visible when you watch an animation of the strange attractor, many of which are now available on the World Wide Web. Disequilibrium can fall into a dynamic equilibrium with a slight shift, and can again be thrown into a new disequilibrium by yet another shift. The strange attractor can be any point within an orbit that appears to pull the entire system toward it.

Chaos theory is based in part on the fact that Newtonian paradigms of predictability do not actually work. Accepting nonlinear systems creates a challenge to scenario planning. Recasting scenario planning to create bespoke futures acknowledges the unpredictability of strange attractors, but hopes to use the process itself (as well as its result) to move the system toward a tipping point. Returning to Barr's diagram, bespoke futures are more like strange attractors than oppositional or political avant-gardist objects.[27] The bespoke futures process can develop attractors to pull the entire system toward new and more hopeful visions of worlds to come. These bespoke futures can become energy fields with which their makers aspire to move the whole system of the world into their orbit.[28]

It would be the height of foolishness to claim that the bespoke futures process will by its nature yield progressive ends. The same methodologies could result in systems of the world that would strike me as far worse than our present. On the other hand, bespoke futures can move us past the Official Futures we all have in our heads (which since 2001, have been permeated by fear), and try to use digital technologies and media design to craft cultural strange attractors—magnets for the imagination that can enmesh their users in a better, more hopeful, and more meaningful set of futures. More and more, meaning itself becomes the central concern. How can we ensure that the phenomenal machines and pervasive infrastructures we have invented—the computers, networks, nifty little portable devices, augmented spaces, interactive entertainments, the list goes on—actually hold as well as develop complexity, rigor, and meaning?[29] Crafting a compelling set of bespoke futures can enable activist positions. Making meaning with these objects and sticky systems will be key to a society that refuses to equate citizens with consumers.

Schwartz writes that "there is a hunger for another set of visions of a possible society," but what is missing in corporate and even nonprofit scenario planning is precisely that vision—that X factor that creates the future. Schwartz discusses the relationship between the unconscious and conscious, and talks about images as the link to the unconscious, yet he does not move to the next step and see the production of images as critical to setting that unconscious free.[30] What the design fields bring to scenario planning is precisely the power to take a discussion and animate it as vision, interaction, and environment. Not only that, bespoke futures engage with the essence of the design process, the crafting of potentialities out of the imagination, and their eventual realization—or at least virtualization—in the world.[31] If there is one thing we ought to be able to do, it is to train a new generation of visionaries, of young people who not only can imagine a better future but can visualize and design it as well.

Finally, I want to stress two things. The word bespoke has a commercial cast to it, and it is exactly that connection to design and the market that I am trying to engage, rather than falling back on the exhausted tropes of oppositional avant-gardism. Second, although I have emphasized aiming high, choosing one future and working toward it through this bespoke process, I want to make it clear that this work would be taking place in a wired world, and one in which these unique shards would add up to multiple sets. I am not talking about a singularity of utopian vision but instead a networked plurality of vision—a plutopia, as I call it—better suited to this century, this millennium.

MaSAI: Massively Synchronous Applications of the Imagination

Participation is a fine thing, but like most fine things, it can be readily commodified. The label "prosumer" (producer-consumer) has been coined for those users of the Web who contribute to commercial culture via uploading without much in the way of commercial recompense. The issue of who owns the content that users generate is not abstract. Many of the major social networking sites have buried deep within their agreements that the sites and their corporate structures have full control over any material uploaded on to them. They even retain the right to continue hosting a page if the person who created the material on it wishes it to be removed (a not-uncommon desire among college seniors who come to understand that their potential bosses have access to their records of youthful exuberance). There are no surefire safeguards against the abuse of prosumerism. One tack is to consciously choose to upload different kinds of media for prosocial rather than prosumer reasons, crafting alternate visions to take advantage of the braiding of simulation and participation.

When large groups of users are able to access powerful universal simulators, what develops is the potential for MaSAI. This massive synchronicity of the imagination is one way to add stickiness to the system. Successive solutions or experiences add layers and outcroppings, which can offer affordances to attract yet more contributions over time. Massive and networked does not always mean faster and more (however oxymoronic that may sound). One of the things that MaSAI can encourage, as we have already seen, is a more deliberate, contemplative approach to this richness, as in the info-triage discussed earlier in chapter 2. This is not to say that to be sticky, a work needs to be collaboratively produced, nor that more collaborators equals stickier work. What I am claiming is that MaSAI opens up a new space for the production and distribution of culture of all kinds.[32]

Here increased participation leads to a network multiplier effect, with intelligences applying themselves to the problem, thereby increasing the likelihood of intriguing solutions and even the generation of new problems. These solutions range from confronting the so-called wicked problems—those most intractable issues of poverty, hunger, and determining what constitutes the good life well lived—to smaller concerns about the individual, household, and even neighborhood or school. The point of MaSAI is to take the maxim of open-source software developers—"Given enough eyeballs, all bugs are shallow"—and apply this to other realms of social life and the built environment.

One project that points the way is called Stardust@home, which has assembled a huge group of people to use the network to search for interstellar dust collected by a recent space mission.[33] In 2004, the Stardust interstellar dust collector passed through the coma of a comet named Wild2 and captured potentially thousands of dust grains in its aerogel collectors. In 2006, the craft returned to Earth and the search for these grains began in earnest. To find these grains—the first contemporary space dust to ever be identified—within the aerogel is no easy task, though, because they are randomly spaced and tiny. It is also difficult to run pattern-recognition software because the traces that the grains leave are similar to other deformations in the gel. Human perception is well suited to this kind of detailed discrimination, however. Twenty-five years ago, the University of California at Berkeley team would have trained a group of laboratory assistants, and set them to work for the next four or five years.

But the Stardust team had another model to draw on. For more than a decade, ordinary people had been not just willing but also eager to turn part of their computer's run cycles over to the Search for Extraterrestrial Intelligence project. The SETI@home distributed computing initiative has been wildly successful for almost a decade. It sends out chunks of data (or "work units") to computers all over the world, and the users then send the results back. Users' willingness to share their untapped computing power means that the Search for Extraterrestrial Intelligence (SETI) project does not need to purchase extra supercomputers or rent time on them. The distribution here is essentially a free gift, and requires little from the participants beyond their signing up, and then allowing the data chunks to flow in and the results to flow out.

Stardust@home moves this process to the next level, asking people to commit to learning how to read the images and then identify the grains. This links users, networks, and machines, and appeals to people's sense of wonderment about space as well as their desire to ferret out mysteries. These communal efforts even appeal to people's vanity, as the most successful participants will be listed as coauthors on any major scientific papers to emerge from the Stardust interstellar dust collector. Stardust@home is wonderful, but it remains a top-down hierarchy. NASA sends up a probe, scientists plan experiments, and distributed users/participants execute the analysis. What about more distributed, spontaneous activities? The next level is to join people together for wider-ranging cultural and scientific processes. These larger-scale, massively parallel applications of the imagination have been manifesting themselves through a whole range of applications, social networks, and cultural initiatives. As discussed in chapter 4, the open-source movement and Creative Commons are two, wildly successful models of this.

R-PR: Really Public Relations

The open-source movement, Creative Commons, and the battles to limit copyright are all weighted on the side of mindful downloading and meaningful uploading, but this Environmental Impact Report on Web n.0 would be insufficient if it stopped there. The culture machine is now proving itself to be one of the key tools in yet another conflict originating in the twentieth century: the power elite's control over public relations versus the rights of the public to know the actual story. Public relations as a means of manipulating the opinions of vast swaths of the population—the manufacturing of consent, as it came to be known—had a number of originators, but none more prominent, nor with a more interesting backstory, than Edward L. Bernays. Bernays, who came to combine insights into the general subconscious with a mastery of mass media, was in fact a nephew of Sigmund Freud himself. After World War I,

Bernays took Freud's pioneering work and retooled the idea of "propaganda" for peacetime use in the world of business, industry, and party politics. Of course, public relations can be seen as an improvement on royal edicts and the propaganda of totalitarian regimes (both of which have the hard power of the state behind them), but its soft power is formidable when it achieves the diffusion that the television afforded it in the second half of the twentieth century. Open-source cultural initiatives offer communities some sort of autonomy of information collection, analysis, curation, and distribution. Add to that the power of distributed computing itself, and then open-source simulation and social modeling becomes possible.

The legacy of spin remains with us to this day, and it would be foolish indeed to deny that the networks and affordances of the culture machine have not actually made it easier than ever to manufacture consent. But at the same time, these new connections and capacities have made the production as well as distribution of what I call "really public relations" or R-PR possible, and can offer a counterbalancing effect. In order to function this way, though, they need to be brought together and aggregated even more rigorously to fight off the unelected, often-unaccountable, corporate interests. As corporations fund more of their own research and consolidate their tactics to bury contrary evidence, open-source networks can function as a countervailing force, obviously in terms of the distribution of research. The first aspect of this can be seen as a twenty-first-century response to the development of public relations, in its twentieth-century form, as a privatized, corporate-controlled manipulation of information.

And it is indeed information that we need, even if we feel we are awash in too much already. Yet how will the information be arranged, distilled, and deployed? These are questions of design. This way of thinking and working looks back to one of the guiding principles of modernist design, which allowed practitioners to serve as conduits and refiners of complex information about social, aesthetic, and scientific systems for mass audiences, thereby adding to the realm of knowledge and democracy. In the 1920s, a remarkable polymath named Otto Neurath exploded into the realm of design. Neurath, trained as a mathematician, was obsessed with universalizing knowledge, working at different times as a philosopher of science, a sociologist, and an economist. His strongest impact by far, however, came with his desire to create modes by which to convey large bodies of information to diverse audiences. Neurath maintained that the twentieth century was generating the huge quantities of data that an educated citizenry required, but did not always (or even frequently) have the capacity to understand. Neurath's innovation was the isotype (the International System of Typographic Picture Education), which we now recognize as those ubiquitous silhouettes indicating men's and women's rooms by means of abstracted figures.

Little description is needed to invoke the almost-hieroglyphic feel of isotypes: road signs, warning labels, and high-voltage indicators.[34] In the words of media philosopher Frank Hartmann, Neurath wanted "to introduce media literacy as enhancing a new form of enlightenment."[35] Neurath used the emergent mass media technologies of his era and created early signposts to the figuration of information itself. His seriousness about the importance of the social sphere translated into a remarkably durable visual iconography. The challenge for contemporary makers and thinkers is to take this kind of social positivism and link it to contemporary networked environments to create a social media that goes beyond sociability into the realms of the useful.

As local and global networks proliferate and intermingle, the web of computer-enabled scenarios will grow ever denser and more interconnected. This will lead to a truly new "new thing" in the world—one capable of contributing to the discussion and development of its own future. Brenda Laurel once noted that "creating interactive simulations of complex systems is one of the most highly leveraged goals we can achieve with our burgeoning technological power. . . . Good simulations will not only help us learn about systems, they may help us evaluate policies and form political goals."[36] This new century brings new problems, as is always the case, but we would be foolish to not take advantage of peer-to-peer networking, file sharing, and massively scaled distributed computing to develop countervailing forces, from a truly populist scenario-building capacity to as yet unimaginable visualizations of change.

Open-source cultural production challenges more than government and corporate centralization; it also serves to empower the citizenry. The more that states and corporations grow, the more individuals need to be able to communicate with others in their own communities and across the globe, if they hope to have any say in their own lives. One of the products of the megastate is the generation of megaquantities of information. Few individuals have the capacity to move through this data, much less fully understand it. But in networks, they can serve as a counterbalance to the "official" take on the information. The Many Eyes Web site is just the sort of intervention that the computer as culture machine makes possible. The goal of Many Eyes "is to 'democratize' visualization and to enable a new social kind of data analysis." The site does so by offering a set of templates, and organizing visual schemata for creating representations of official and personal data sets.[37] Describing Many Eyes with this language may make it seem like a site that only a statistician could love. Once people start to explore how changing the visualizations of the same data can produce entirely different reactions, the power of such tools becomes evident.

We can, in fact, use these culture machines recursively, deploying them to analyze ourselves—MaSAI in the service of looking at our own intelligence and imagination. Computers are the best devices that we have ever created to generate and evaluate probabilities. They are ideally suited to develop simulations of potential outcomes; they are scenario builders. In a world where image, sound, and even interactivity become central to argumentation, the computer's ever more sophisticated visualization technologies become key. A sensibility that deploys the culture machine to create visions of the future that we actually want to live in can keep us from slipping back into the narcotizing state of entertainment for which television conditions us.

Plutopian Meliorism

The culture machine's melding of techniques and media opens rather than closes communicative potentials. The expansiveness of the culture machine augurs pluralism, which can strike some as open and embracing, and others as uncommitted and market driven. At its best, pluralism in contemporary culture will lead us toward a concept that I introduced earlier as plutopian. This neologism tries to save the hopefulness and striving for a better future inherent in the word utopia while acknowledging that one size does not fit all, and that a blending or hybrid of utopian thoughts and practices is the best that we can achieve.

What these visions of the future can offer is a way out of the banalities of the "marketplace of ideas," and into one of competing or plural utopias. Plutopias are profoundly American in concept—deeply ingrained in a culture that promises not happiness itself but rather the pursuit of happiness as a founding principle. To promise happiness is akin to offering a single utopia, a uniform vision of satisfaction. The pursuit of happiness, on the other hand, is inherently about unfinish.

This quest is a process, not an end point, open to a range of outcomes. This openness can only thrive in an environment where we have waded into the secret war on the right side. Ensuring the capacity to upload, and doing so in a meaningful way, is less a revolutionary strategy than a pragmatic one.

If we look back a hundred years, we can invoke the pragmatic philosophers, and even more centrally, their concept of meliorism. William James defined meliorism as "an attitude in human affairs" rather than a creed: "Meliorism treats salvation as neither inevitable nor impossible. It treats it as a possibility, which becomes more and more of a probability the more numerous the actual conditions of salvation become."[38] Meliorism's attitude toward the possibility of improvements melds well with the cultures of the computer. Television does not improve so much as metastasize, growing to gargantuan, home-theater size in the den, spreading out to multiple incarnations in every member of the family's bedrooms, into our cars, on to our personal digital assistants, and into ultrabright "outdoor models," recently reserved for the ultrarich but soon to be in every backyard space near you.

The computer's trajectory, on the other hand, strikes me as hopeful and ever upward, from 1.0 to 2.0 to n.0 to an asymptote of infinity. Of course, some of this is just the hype of new releases and unnecessary upgrades, but even short-term history tends to smooth these jagged edges off the upward-tending curve. The quantitative increases in speed, sophistication, ubiquity, mobility, miniaturization, and personalization become, or at least have the capacity to become, qualitative changes in the ways that we make culture. Let us not forget that the word culture derives from the same root as the words cultivation and agriculture, so to speak of the culture machine as growing and evolving through encouraging uploading is no oxymoron.

To weigh in on the side of uploading is a modest goal, not the perfection of utopianism, but the pragmatics of plutopian meliorism. Because meliorism takes as its goal making things better through concerted effort, it is a habit of mind and a mode of practice that aims for realistic optimism, rather than passivity, pessimism, or nihilism.[39] John Dewey wrote that "the striving to make stability of meaning prevail over the instability of events is the main task of intelligent human effort."[40] What could be more melioristic than mindful reception and meaningful production, even if these exact phrases were hardly a part of Dewey's lexicon?

Proposing plutopian meliorism as the ends and the culture machine's aesthetics as the means does not prescribe unitary intent, but exactly the opposite—opening the possibilities for unfinish. You cannot, nor should you want to, close down avenues to cultural participation or pleasure. That is a choice, but unlike television, it will not be the only choice if the affordances are built in, and if others model different behaviors (early and often, as they say).

Enlightenment Electrified

If cultural diabetes is as big a problem as I claimed, the issues I have been discussing in this book are less cures than preventive measures, with an emphasis on curbing consumption and concentrating on the powers of production, positing that material progress is relatively uninspiring without an incremental increase in the creative potential of the individual and society together. What makes it a remix, rather than a revival of Enlightenment rationalism, is that the twentieth century's great accomplishments—antisexism, antiracism, and anticlassicism—will be soldered into the equation. In other words, and to add to our transtemporal mix, a new Enlightenment, electrified.

The Enlightenment is a label we retroactively applied to a braided strand of transformations, events, and personalities stretched out over multiple decades, if not centuries. The present moment of unimodernism, if it deserves to be seen as an Enlightenment Electrified, must be even more pluralistic in its approaches and manifestations, as befits the iterative power and play of the culture machine. This very pluralism dismays programmatic critics who miss the sureties of high modernism, much less those of a regimented, patriarchal, religious society. But the certitude of those moments is gone.

The Enlightenment's metanarratives—of universal human rights, progress through inquiry, the rightness of reason, and the ascendance of the secular—were all attacked in the second half of the twentieth century by forces from both ends of the political spectrum. The far Left mounted a critique of the universality of these claims and the truth-value of truth itself. Variously gathered together as poststructuralism, deconstructionism, and postmodernism, these were often thoughtful critiques of a hegemonic system. With distance, they can be seen as extensions of rather than a complete disavowal of the Enlightenment inheritance. The critique was fueled by outrage that the enlightenment of some could so easily coexist with the economic, colonial, racist, and sexist oppression of others.

The climate that followed the 89/11 period brought into question, for at least some of us in the rich West, the lucky North, a sense that whether we admitted it or not, one of the appeals of the post-'89 period was precisely the ahistorical fantasy that we were beyond history itself. That fantasy was laid to rest with 9/11 in New York, 3/11 in Madrid, and 7/7 in London.[41] The contemporary moment must confront its adversaries. I disagree with French sociologist of science Bruno Latour's contention that we have never been modern. The issue in Latour's denial is "we," as we, of course, are rarely anything jointly. I do maintain that "we" are all at least partially unimodern. The unimodern is like William Gibson's aphorism about the future—all around us, but unevenly distributed. The computer is a machine that both figures individuals, and allows them to more easily coalesce into social and market forces. The computer is a rational device par excellence, driven by the exigencies of the Enlightenment, but it is also a desiring machine of the new economic order, a tool that demands new tools, that makes possible individuation and nichification. Like all desiring machines, the computer has a part to play in the large-scale conflicts that convulse our culture, including the balance between the sacred and secular.

The question becomes how to compete with the devotions and obligations of the sacred with nostrums about collaging digital snapshots in Photoshop or fabbing WYMIWYM knickknacks. To answer this, I would recall that the driving, burning issues of high modernism have burned themselves out. Flatness achieved, tonality totaled, the building definitively stripped of its decoration and turned into a box containing people, and the new novel denuded of character and plot—twentieth-century modernism was almost done in by its own success. The running room that modernism created was in turn commodified over the course of the twentieth century, eliminating the space for continued dreaming and creativity. The lack of running room between cultural production and its commodification, first into niche, then into kitsch, has disappeared. The heroics of modernism, driven by its own lusts for ground zero, set a higher bar for unimodernism, but that is the challenge for us: to use the computer, which decreases classic notions of running room even further, to reinvent it for the twenty-first century.

Revolution has been co-opted by the marketers, on the one side, and fundamentalist death cults, on the other. Technological determinism is not the answer. Technologies certainly open up spaces, but they also close them down. It is hardly as simplistic as the idea that "you can't stop progress." Technologies are supported, deployed, and abandoned for a huge, interlocking number of reasons. Some have to do with the technologies' innate qualities, yet many of them have to do with the economics of dispersal and marketing, timing and political will. What is interesting for the unimodern moment, though, is how technologies have been rolled out with an unprecedented "presentness." They are introduced as social or commercial "revolutions" without being slotted into an overarching narrative of general progress.

Striking a balance between downloading and uploading, practicing info-triage, seeking out and adding stickiness, and crafting bespoke futures are all important in their own right, even if the stakes were not so high. But at this moment, the stakes are high indeed. One of the problems of any kind of cultural conservatism—even one that attempts to preserve the spirit of the modern—is its base in nostalgia. It seeks to return to an imagined past rather than imagine a future it wants to inhabit. So if the way back is blocked, what happens to those feelings? Where do they go at this point?

I have already made the claim that the end of modernism's capacity to inspire created a hole in our society's heart. That hole creates a vacuum, and a vacuum must be filled, whether in nature or culture. What flowed in first was the mindless download, but what has lately been competing with that is the religious impulse that the project of the Enlightenment struggled for so long to push out and then hold back. The Enlightenment never fully did away with the religious impulse, and modernism, for all its neo-Nietzschean self-satisfaction about the so-called death of god, did not do so either. The notion of separate spheres for culture and religion, and science and religion, were stronger in the era of high modernism than they are in our own, unimodern moment. This is not to say that religion was under attack but rather that the secular had a great deal of confidence in its own expertise, pleasures, and autonomy. As no less a luminary than Charles Darwin once said, "I do not attack Moses, and I think Moses can take care of himself." What the Enlightenment and its most recent, modern stage did manage to do was create a vital, vibrant, and meaningful secular culture.

We have to return to the central questions: What replaces the kinds of cultural aspirations and energy that the avant-garde inspired in its makers and admirers? How can the culture machine fill the void that entertainment drills out of our psyches? It is that voiding of meaning that allows the old dark forces to smuggle themselves back in. While this may sound irredeemably negative about spirituality, there is some space for negotiation. After all, before his death, even famed geneticist Stephen Jay Gould attempted to reconcile science and religion by granting each its own sphere of influence.[42]

The reason that the computer is an ideal tool to fight for the secular is because innovation is the machine's first, melioristic principle. The key is to keep secular culture from being seen as either entirely technicist—the reduction of the scientific to mechanics—or as entertainment—the reduction of culture to distraction. The most virulent warriors against innovation are those fundamentalists who see themselves as the implacable enemies of reason in its broadest sense. They seek less a return to a glorious past than to an antirational religious paradise here on Earth. They claim to loathe the culture of entertainment that modernity so successfully promotes, yet they use the same media in their own campaigns: television, mobile phones, and the World Wide Web are central to their organizational, recruitment, and even paramilitary activities. Fanatics dream of using computer networks to enable decentralized groups to set off atomic explosions, which would be captured by television, and like the fall of the Twin Towers as well as the carnage on the Madrid trains and London subways, be broadcast live to the world.

The decline in modernism's capacity to inspire has not gone unnoticed. The former head of Al-Muhajiroun, a radical group based in London and a suspected member of Al-Qaeda, Omar Bakri Muhammad, holds virulently anticosmopolitan views, including the assertion that "the life of a nonbeliever has no value." His moral nullity is self-evident, but it is his contempt for the culture in which he found himself after fleeing repression in his native Syria that is of interest in this case. Until recently, Bakri Muhammad lived in England and took it on himself to call for the ultimate conversion of the British Isles to Islam. Before he was expelled from the United Kingdom for subversive activities, Bakri Muhammad was asked why he was so certain that "the black flag of Islam" would soon fly from Number 10 Downing Street. His contemptuous response was because "Western culture is nothing more than entertainment," and this entertainment "has no answers about the meaning of life and death, which is life's biggest challenge."[43] Bakri Muhammad is himself an indefensible thug, and this was warmongering against not just the culture of entertainment but also the country that for more than two decades had sheltered him from political enemies in the Middle East. But his denunciation of the West needs to be answered with more than either the flitting attention of the media itself (since his move from London to Lebanon, who even remembers him?) or with theological disputation (either from more moderate positions within his particular faith or from other traditions).

How are we to develop content and meaning so that the messages are worthy of their media? The twentieth-century conflicts between highbrow and lowbrow culture ground themselves to a halt, leaving the space for a hypercommercialized middlebrow to step over the exhausted combatants and receive its laurels in high definition on television. There is a history to this reversal of fortune, and it is worth revisiting two of the combatants decades later, in retrospective moods. Before the *New Yorker*'s longtime film writer Pauline Kael died in 2001, she said that when she was attacking high culture in defense of low culture, she had no idea that high culture would eventually disappear. In the last decade before her death in 2004, Susan Sontag was surprised that in the end, she had to defend the very idea of seriousness. One thing that bound these two otherwise-different voices of 1960s' cultural criticism together four decades later was exactly that sense of amazement at the conclusions of their careers—that the shibboleths of their youth had in fact given up the ghost, that the gods of high culture could retire just like Zeus and Hera (but without an Olympus in which to live out their eternities), that in winning the battle, modernism (and even more, postmodernism) lost the war. Writing a quarter century ago, Sontag observed that "stripped of its heroic stature, of its claims as an adversary sensibility, modernism has proved acutely compatible with the ethos of an advanced consumer society."[44]

Without knowing it, Bakri Muhammad, the nihilistic imam, repeated Sontag's critique in his own language. The "advanced consumer society" that modernism proved so "acutely compatible with" is precisely the culture that television enabled, promoted, and made inescapable. Certainly Bakri Muhammad sees religious piety as the "cure" for this culture of entertainment, but so do Hasids in the Israeli settler movement, ultranationalist Hindu fundamentalists in India, and Christian theocrats in the United States. Instead of calls to bloody jihad, this last group of blow-dried preachers cloak their dreams of rapture in the language of polls and focus groups, signaling their intent to the righteous through code phrases like "purpose driven," "faith based," and "family values."

As a nonbeliever, I am compelled to defend my views of secular culture, but dismissing Western culture of the twenty-first century as entertainment is something I have thought myself in bad moments. We must ensure that the secular culture of the West is powerful enough to stand up to sympathetic critiques from within, not to mention from theocratic thugs, wherever they are from. This is not a "crisis," a word so overtaxed by its deployment in academic paper titles and political attack ads as to be almost meaningless, but rather a chance to survey our culture and take stock of its best attributes—a redemptive form of criticism.

We need answers that look forward into the potentialities of the twenty-first century instead of reflexively backward, as though our only hope is a retreat to hazy highlights of a Western civilization survey course. We need to embrace the idea that what can be most interesting about a particular cultural moment are those things that could not be done before by earlier generations and their technologies.

To strengthen serious secular culture and its immune system to ward off these attacks, you have to create affordances for both production and consumption, seeing the whole of culture as a process as opposed to a result. But the essence of the digital, as noted earlier, is an aesthetic of unfinish, an understanding both literal (the work is never out of beta) and metaphoric (the digital is always in flux between points, flitting about and flickering on the grid). In the nineteenth century, the French author Paul Valéry famously said, "A poem is never finished, only abandoned," and in the era of unimodern unimedia, we all become poets after a sort.

The very success of the Enlightenment project in transforming the culture of the West engendered a backlash both from within and outside. But we who disagree with the rising tide of theocrats have an ally. The computer as culture machine can empower those of us who believe in the importance of plural rather than unitary meanings and secular rather than sacred intents. The computer was not designed with this goal in mind, but then again, neither was the printing press. Gutenberg invented his press to manufacture Bibles for the burgeoning middle-class market, but his innovation allowed others to produce, distribute, and consume their own pluralistic as well as secular visions. Without the development of the press and mass printing, it is doubtful that the Enlightenment would have been as powerful or widespread as it was.[45]

Governments certainly use computing and networks as bulwarks against terror: closed-circuit television surveillance, massive data-mining operations, and the technological gewgaws that once seemed to be restricted to James Bond fantasies but are now part of our daily lives. Yet as important and often ham-fistedly used as these systems are, it will also be the ways in which the rest of us use computers and networks as our culture machines that will make the difference in terms of creating a society eminently worth defending. I prefer for at least the near future to develop utilities rather than promulgate manifestos. Utilities add functionality to software systems. They can build on themselves, or be built on in turn by further utilities and developers. Utilities are meliorist rather than revolutionary, unlikely to call for the destruction of a system in order to save it. There is an inherent modesty to the utility, an additive improvement and immediate functionality. Much of this book can be read as a series of interlocking, self-reinforcing utilities meant to increase the abilities of the culture machine to create meaning.

If anything, the computer allows the human creative spirit even more flexibility and greater potential than the printing press because it synthesizes so many other media forms. Educational theorist Howard Gardner proposed a theory of multiple intelligences more than a quarter century ago, writing that in addition to the linguistic and mathematical aptitudes that most standardized tests measure, people can have spatial, musical, or even intrapersonal intelligences that go untracked. Whether or not one agrees with his underlying thesis, Gardner's emphasis on multiplicity melds well with the era of the culture machine, which allows for the expression of these aptitudes on newly global scales.[46] The culture machine's very novelty, its capacity to generate new forms and new explorations, can save us from turning to nostalgia, from looking for the hope for the future in a return to the past. We cannot go back, neither to those few short decades of heroic modernism, nor to the centuries of traditional faith in mosques, churches, and temples. While claiming to move forward, both utopian modernism and millennial religiosity are in fact locked in stasis. I would argue instead for the workaday realities of meliorism as the hope for a regenerated secular culture.

Identifying the best uses of the culture machine is not simply an academic exercise. In creating and perfecting the computer, we came to rival Gutenberg and his press. It is not (old) New Economy hype to say that in myriad ways, the computer will have more of an impact on culture than the printing press did. But we need to nurture the culture machine's nascent capacities and not lock ourselves into a series of downloading presets just because we can imagine them ourselves, or because there is a momentary economic or political rationale for them.

If those who had followed in Gutenberg's wake used the printing press to simply manufacture Bibles for the middle-class market, rather than inventing the newspaper, pamphlet, almanac, poster, novel, and even comic book, we would have judged them harshly. In our case, it is even more difficult to make sure that the future will be able to invent for itself, because so many of our systems are designed with backward compatibility in mind, with grandfathering in older systems. Again, this is worthy, and it is important to ensure that we do not abandon what we have already made and digitized with every change of system, but it also means that the decisions we are making now (or for the vast majority of us, not making, because it is usually only a tiny cadre of programmers who make these decisions, on the fly, and under the pressure of delivery dates and quarterly reports) will have a direct—if impossible to quantify—effect on our descendants. If as a culture we have slowly swung around to considering what our impact on the ecological environment will bequeath to our children, grandchildren, and beyond, it is now time to do that for our technological environments as well. We can and must leave a better legacy than this.[47]

PATRIARCHS

PLUTOCRATS

AQUARIANS

HUSTLERS

HOSTS

SEARCHERS

GENERATIONS

GENERATIONS

HOW THE COMPUTER BECAME OUR CULTURE MACHINE

▶ notes: pp. 194–197

Multiple Histories, Multiple Choices

The secret war between downloading and uploading, unimodernism, info-triage, power and play, informationalism, and sticky media are twenty-first-century issues, but they have a rich and deep past. The computer and its networks are the most complicated machines that humans have ever constructed. Their history is not an easy one to record, tying together as it does abstruse mathematics, complex engineering, fortunes made and lost, and a rigorous accounting of the computer's transformative powers over how we live, work, and play. Yet we require an understanding of how the computer developed as it did if we are to understand the arguments throughout the rest of the book. This means that we have to be ruthlessly selective in constructing a history to suit our purposes, choosing comprehensibility over comprehensiveness.

In the sections that follow, I trace the ups and down of simulation and participation through six discrete generations of computing. I offer a polemical account about the development of an ideal of computing, tracking what the evolutionary biologist Richard Dawkins calls "memes," the intellectual equivalent of genes or self-replicating ideas.[1] The pages that follow are filled with narratives about people who become infected with memes about the computer as an essential culture machine. The two memes central to the development of the computer as a culture machine are simulation and participation, and tracking their interplay is key to developing these historical narratives. These memes usually preceded the technology, and those infected then spent years developing computers and systems to bring these essences into existence, with each generation building on or challenging the last.

I characterize the first generation as the patriarchs—here represented by the idiosyncratic visionary/bureaucrat/scientists

Vannevar Bush and J.C.R. Licklider—who establish the founding memes in the early years after World War II through the early 1960s. They are followed in turn by the Plutocrats—Thomas Watson Sr. and Thomas Watson Jr. of IBM—who make a business out of computing, centralizing the operations into top-down bureaucracies during the 1950s and 1960s. In reaction to the buttoned-down, all-business attitudes of the Plutocrats, the Aquarians of the 1960s and 1970s—people like Douglas Englebart and Alan Kay—expand on the more open-ended ideas of the Patriarchs, and develop the paradigm of visual, personalized, networked computing. In the 1980s and 1990s, the Hustlers—Microsoft's Bill Gates and Apple's Steve Jobs—commodify this personalized vision, putting a distinctive, "new economy" stamp on computing. Building on the installed base of all these users as the new millennium looms, the Hosts—World Wide Web inventor Tim Berners-Lee and open-source guru Linus Torvalds—link these disparate personal machines into a huge web, concentrating on communication as much as technology, pushing participation to the next level. The sixth generation, that of the Searchers—named after but hardly limited to Larry Page and Sergey Brin of Google, the search algorithm that became a company and then a verb—aggregated so much information and so many experiences that they rendered simulation and participation ubiquitous.

There are three default ways of telling the history of computing, and the interesting thing is that people rarely tend to blend the narratives. There is the technical and scientific history of computing, which is frankly the least understood and disseminated. This is the story of algorithms and circuit diagrams, tubes that became transistors, and laboratories at universities like MIT and Stanford as well as companies like IBM and AT&T. Without this work, there would be no software, no computers at all, but the details are technical, and those involved are far from the spotlight. This is a technography—

a written record of the technology composed, in the main, for experts.

Then there is the story that most people know. That is the story about fortunes made and lost. During and after the great Internet bubble of the 1990s, there were instant history machines for the so-called new economy—magazines like *Business 2.0* and *Fast Company*—that reported on the ups and downs of the geek gods and their "wealth creation." Here are the tales of Microsoft stock bought at twenty dollars and sold at two thousand, Bill Hewlett and Dave Packard working in their rented Palo Alto garage, Ross Perot quitting IBM to found Computer Data Systems in Texas, Jeff Bezos opening an online bookstore, naming it after the largest river in the world, and then getting on the cover of *Time* magazine as the CEO of Amazon.com, and Mark Zuckerberg transforming the Harvard University first-year-student listing service into Facebook, the dominant and most valuable social media site in the world. These are the stories that have sustained the bulk of people's interests in the history of computing. This is the history of computing as plutography, stories about money.

There is another small but growing strain that locates the transformations of our world in the work of computing's visionaries. As far back as Howard Rheingold's *Tools for Thought* written in the mid-1980s, there has been an alternative narrative featuring people like the irrepressible hypertext impresario Ted Nelson and even drug guru turned cyberpundit Timothy Leary—an intellectual's history of computing.[2] For the scholars studying hypertext poetry, the students in new media departments, and those with a cultural interest in computing, these are stories of secular saints, a hagiography of sorts. To get to a workable understanding of the history of the culture machine, we need to braid these three strands, looking at programmers, millionaires, and dreamers. That these strands can all combine in the story of one person, one machine, or even one company is all to the good.

The Warriors: A Prehistory

Life was simple before World War II. After that, we had systems.
—Rear Admiral Grace Hopper

The question to begin with is not, "*What* is a computer?," but rather, "*Who* is a computer?," because computers were humans first and machines second. Computers were people, usually women, who computed numbers, tabulated results, and published lists or matrices with the results. They worked for nineteenth- and early twentieth-century businesses and government ministries, and laid the groundwork for the data-driven, statistically charted, numerically marked world in which we now live. The demand for information during World War II pushed the human computers past their capacity to produce what was needed for the war effort, and new systems were required.

The military acronym C^3I stands for "command, control, communications, and intelligence." Twentieth-century warfare proved over and over again that the side with superior C^3I wins. Unlike World War I, which was primarily a European conflict that spilled over into some of its immediate neighbors in the Middle East and North Africa (and involving the participation of former British colonies like the United States, Canada, and Australia as allies), World War II really was a global conflict, with battles not just in Europe but also across Africa, the Soviet Union, China, Japan, vast areas of the Pacific, and even reaching the United States with the bombing of Pearl Harbor. It was a war fought across twenty-four time zones, on land, by sea, under water, and in the air, and concluded with the most fearsome, technologically complex weapon ever invented: the atomic bomb.

Integrating all the information to wage World War II was simply beyond the capacity of human computers. There was too much incoming data, and the demand for accurate, formatted, and timely information was literally one of life and death. The first order of business in the development of computers as machines was the simulation of computers as people. The U.S. government supported the development of mechanical means to do what the human computers had been doing up to that point. The driving need here was for ballistic tables. A large-caliber gun or missile launcher is affected by weather conditions, the projectile's charge, the target's distance, the gun's elevation, and dozens of other factors. If the war effort was to extend its C^3I and dominate the battlefield, each different kind of gun had to have different ballistic charts calculated for it.

The success of the C^3I initiatives were obviously the military's first and foremost objectives during the war, and if this meant replacing the "who" with the "what" of computing, so be it. But there were a few key people during the conflict who saw that the powers of computing, if spread wider than the laboratory and the war room, would be a huge benefit to humanity. By moving toward the goal of participation and melding it to simulation, they were able to shift the focus from the "what" back to the "who" again. These were the Patriarchs.

The Patriarchs: Vannevar Bush and J.C.R. Licklider

The world has arrived at an age of cheap complex devices of great reliability; and something is bound to come of it.
—Vannevar Bush

People tend to overestimate what can be done in one year and underestimate what can be done in five to ten years.
—J.C.R. Licklider

There are many mathematicians, early computer scientists, and engineers who deserve to be considered part of the first generation of pioneering Patriarchs. They include Alan Turing, already discussed in chapter 2; mathematician and quantum theorist John von Neumann; cyberneticist Norbert Wiener; information theorist Claude Shannon; and computer architects like the German Konrad Zuse, and Americans J. Presper Eckert and John Mauchly, who developed ENIAC, the room-sized machine at the University of Pennsylvania that we recognize as the first general-purpose electronic computer. These were the Patriarchs who set the parameters for computer science, laying out the issues for software development, building the original architectures for hardware, and creating the cultures of computer science and engineering. They deserve bookshelves of coverage, and the past decade or so has seen an explosion of publishing about them, with biographies and critical assessments coming out on many major figures. Here, though, I concentrate on just two: Vannevar Bush and J.C.R. Licklider.

Bush's biography is subtitled *Engineer of the American Century*, and it is not a bad estimate of the influence that Bush had on the culture and organization of scientific research in the second half of the American century. During World War II, President Franklin D. Roosevelt appointed Bush as his director of scientific research and development, and in that position Bush radically changed the culture at not only MIT, where he had been vice president and dean of engineering, but also other major research universities across the country. Bush solidified the intertwining of universities and the military during and after the war. Prior to his efforts, research and development into war-making technologies were housed in military institutions. After Bush's concerted push, these efforts shifted to laboratories based on academic campuses.[3]

The military would be guaranteed the research infrastructure necessary for waging technologically and informationally complex air and ballistic missile conflicts, and the public, at least in theory, would benefit from a steady stream of support for basic scientific research and associated, nonmilitary, spin-off technologies.[4] The spin-offs from these military-funded research and development university projects spurred the development of Route 128 outside Boston (home to companies like Raytheon and the Digital Equipment Corporation), which was in turn the model for Silicon Valley (where Stanford's engineering and computer science departments led to everything from Hewlett-Packard to Intel to Yahoo! to Google).

But all of this intertwining of knowledge and destruction had an impact on the Patriarchs' thinking. Bush in particular wanted to use the new technologies to improve our capacity to process and understand information.[5] In "As We May Think," an article published in the *Atlantic Monthly* just after the war ended in 1945, Bush wrote about an imaginary machine that he called the Memex, which simulated the human brain's capacities for associative thinking.[6] Bush's article remains an amazing read all these years later. It offers a capsule history of computing, or rather precomputing, discussing Gottfried Leibniz's mechanical calculators and Charles Babbage's difference engine, and explores information overload (remember this was 1945), imagining a fully multimediated "future investigator" wearing recording technologies that date, time stamp, and synchronize both research findings and observations. But it was Bush's detailed description of the Memex—short for "memory extender"—that serves as an original meme for the computer as culture machine.

The Memex proposed using microfilm technology to allow its users to track down what they were looking for and then create associative links between these information nodes, both for their own use and the use of others. This attention to the associative mode of meaning making has influenced the designers of hyperlinking systems for more than half a century. That Bush was proposing an analog, mechanical device using chemical photography as its basis was not limiting for either Bush or those who followed him. As Bush wrote so presciently, "It would be a brave man who would predict that such a process will always remain clumsy, slow, and faulty in detail." That the Memex was based in technologies that might never have actually worked, and that digital rather than analog computing turned out to be the "real" future, is irrelevant. What is important for our meme quest is that Bush is considered the foundational figure for hypertext, and his work inspired generations of computer scientists.

It is also important to see Bush's promotion of the Memex as a way to transform the death-dealing technologies of World War II into something beneficial to all of humankind, and something that expanded the power of these ideas beyond the tiny technical and military communities that were then using them. Like J. Robert Oppenheimer, director of the Manhattan Project, Bush was part of a generation of scientists who were more heavily involved in the development of weapons of mass destruction than any other previous one to them. Albert Einstein, a generation older, stood at an Olympian distance from the destruction caused by atomic energy, becoming a secular saint of genius, while Oppenheimer, who had been one of the architects of the bomb, became increasingly despondent about the human capacity for self-destruction.

Bush, like Oppenheimer, had been actively involved in the war effort. He had been instrumental in the strategic planning for the firebombing of Japan and the overall architecture of the assault on Axis industrial sites. In other words, for month after month, Bush had been applying everything that he knew about systems theory to the utter destruction of human beings and their works.[7] That his first public statement after the war should be about the Memex is a stirring example of the Patriarchs's dichotomies.

The second Patriarch, J.C.R. Licklider, though barely remembered today, not only contributed to the dream of simulation and participation in his own work, he also dedicated himself to nurturing the work of others toward this dream.[8] He drew much from his relationship with Bush, who had been the most politically powerful Patriarch. Licklider, whom everyone called Lick, was an experimental psychologist working with sound when he started out at MIT in the 1930s. Like so many of his peers, he joined the war effort, in his case concentrating on sonar, which led in turn to an interest in large-scale data analysis. This piqued his interest in computers, and in particular, he became obsessed with the question of why computers could not be more like other experimental apparatuses, which react in real time to the researcher's inputs. In other words, Licklider was looking for a simulation of the relationship that a biologist has with a microscope or an artist has with a palette: he envisioned the computer functioning in partnership with its user; indeed, one of his most famous papers was titled "Man-Computer Symbiosis" (1960).[9] This was a radical notion in an era of slow machines.

What this symbiosis implied to Licklider was that users needed not just a personal relationship to their machines but personal computers as well. This was a radical stance in the 1940s and 1950s, when room-sized mainframe computers were the norm, and the idea of time-sharing—that is to say, different users submitting their problems to the mainframe to be worked out simultaneously—was considered far out. Licklider's vision of participation went beyond time-sharing on large mainframes and evolved into what was then a fringe concept: that each person would have their own computer. Like so many of the Patriarch's most innovative visions, this one took decades to come to fruition.

Similar to his mentor Bush, Licklider balanced his time in the lab with government service. During his tenure as a high-level administrator at the U.S. Department of Defense Advanced Research Projects Agency (ARPA), Licklider was charged with funding a wide array of projects relating to military applications. But he made sure that within this role, he was also funneling support and securing government funding for those who formed the core of the Aquarian generation in the 1960s and 1970s. This later dedication grew directly out of Licklider's own research. Once Licklider concentrated on the symbiotic communication between human and machine, he began to extend his thinking to include communication between the machines and by extension the users. In 1962, he only half jokingly referred to this concept as the "Intergalactic Computer Network." Licklider was thinking through the overall cultural implications of the computer if it simulated the human capacities of communication and symbiotic relationships with other media. We will return to these themes with the Aquarians, but before that, there are the Plutocrats to contend with.

The Plutocrats: Two Men Named Thomas J. Watson

Think.
—Thomas J. Watson Sr.

Our future is unlimited.
—Thomas J. Watson Jr.

There are few companies that have been around since the nineteenth century, fewer still that have been at the forefront of information technology for most of that time, and only a handful that are capitalized in the billions of dollars. The International Business Machines Corporation, better known as IBM, is all of this and more.[10] We started by talking about the Patriarchs and their visions, but it is vital to realize that for decades, what general public consciousness about computing there was bore little of the Patriarchs' stamp. If you were to ask people in 1959 who the important players in computing's decidedly short history were, they would focus more on the commercial companies that were supplying computers to business, and in doing so, changing the way that the United States and the rest of the world was to do business. These firms included companies like Digital Equipment Corporation (DEC) and Hewlett-Packard (HP), but worldwide, above all others, there were the blue-suited legions of IBM. And they would think of one name, shared between two men: Thomas J. Watson, senior and junior. The elder had built the world's most formidable mechanical data processing company; the younger ensured that its dominance continued in the digital era. Between them, they established the commercial practice and culture of computing for decades.[11]

IBM resulted from the merger of a group of companies (the oldest of which was founded in 1888) that produced the machines that rationalized, streamlined, and quantified U.S. and then global business. It manufactured scales, time cards, and most famously, punch card tabulators that allowed for the storage and analysis of ever-larger amounts of manufacturing, distribution, and sales data. By 1915, a young salesman named Thomas J. Watson had risen from a regional office to take over the company, then named the Computer Tabulating Recording Corporation. Within a decade, he had changed the name of the company to the International Business Machines Corporation, and encouraged the use of its acronym, IBM. Watson Sr.'s famous motto was "Think," but the driving ethos of the company was "Sell." In an era when the art of selling was revered as being next to godliness, Watson was known as the world's greatest salesman (as well as being one of the world's richest people).[12] Watson Sr. understood that IBM had to keep its eye on computers and help to shape their future. During the Second World War, the company cofunded the development of Harvard University's Mark I computer, and IBM scientists and engineers established and solidified linkages with the military that became increasingly critical to the company in the postwar era.

By the time that Watson's son, Thomas Watson Jr., was positioned to take over the company in the early 1950s, the biggest question they both faced was how to confront the changes that digital technologies would have on their company. Should they invest in digital computing, or would this undercut the profits on their flagship mechanical calculators and paper card tabulating machines? Watson Sr.'s contribution was essentially to hand over the company to his son at the very moment when this decision became central to IBM's fortunes. In so doing, Watson Sr. ensured that the computer would make it out of the laboratory and into businesses worldwide, with the huge infrastructure of sales and support that IBM had already built up during the prewar years and the postwar boom.

It was to Thomas J. Watson Jr.'s credit that he negotiated the smooth transition between the two regimes at IBM. In the 1950s, he managed the development and release of the 650 series of computers, which was the first major commercial computing endeavor. He made sure that they were engineered to take their input from IBM's already-established base of punch card processors. Doing this ensured not only backward compatibility but also helped to lock in IBM's consumer base as the company moved itself and its customers from the mechanical to the electronic era. The Plutocrats, whose generation made a business out of computing, often faced hard choices between the technological innovations that their engineers could produce and the willingness (or unwillingness) of their customers to adapt to these changes. With the 650 series decision, IBM made sure to grandfather in as much of its preexisting data processing infrastructure as it could. A decade later, Watson Jr. broke with this model, determining that the architectures of the computers IBM had been selling throughout the 1950s were not going to be robust enough to handle the next generation of needs and capacities.

Watson invested a huge amount of personnel and financial resources in the System/360 series, which was not backward compatible—meaning that he was taking a huge risk. *Fortune* put him on its cover with the headline "The $5 Billion Gamble," questioning whether his customers would follow him into the next step of computing as he saw it. His customers did follow Watson's lead, however, and the System/360 is considered not only one of the biggest gambles in business history but also one of the best. So powerfully did his plan work that within a decade, IBM was being sued by the U.S. Department of Justice for antitrust violations, and Big Blue was often reviled as a soulless, technocratic monster. Mainframes, which came to be dominated by familiar names like DEC, Honeywell, and most important, IBM, were centrally organized and hierarchical, maintained and controlled by an elite corps of data processing professionals—midcentury mandarins—servicing business, military, and scientific clientele. Data entry was commonly done through punch cards imprinted with the warning "Do Not Fold, Spindle, or Mutilate"—good advice but alienating for users. That era's machines were alphanumeric, with no visual computing, and the barest minimum of give-and-take between users and machines. Most people experienced computers through intermediaries, and the software was often entirely opaque to everyone except the programmers and input operators.

Other Plutocrats include Gordon Moore, one of the cofounders of the dominant chip maker, Intel. His insistence that computer-processing power would double and the price would half every eighteen months became known as Moore's law, and became another of the crucial memes for commercial computing.[13] Moore's law means that processing heavy, overly expensive concepts you develop today will still be feasible next year when the power goes up and the price comes down. Moore was to establish another of the memes that the Plutocrats contributed to the culture of computing: the spin-off. Moore and Philip Noyce, another of the Intel cofounders, were both members of the famous "Traitorous Eight," who quit working with the increasingly unstable William Shockley at the pioneering Shockley Semiconductor Laboratory to set up their own, competing company, Fairchild Semiconductors. Moore and Noyce spun off again to found Intel, which has inspired literally tens of thousands of entrepreneurs with dreams of establishing their own computer companies and becoming the Plutocrats of their generation—a variation of the participation meme.

Even though IBM scientists won multiple Nobel Prizes and Turing Awards in the heyday of Big Blue, the reason that we look to the Watsons is that they took the dreams of the Patriarchs and turned them into marketable products. What marked the era of the Plutocrats was a stolid reliability, the quarterly projections of the sales manager rather than the vision of the inventor. Their contribution to the meme of simulation was simple: they wanted computers to mimic other, commodifiable products, much like the Watsons' mixture of ensuring customer buy ins, and the occasional willingness to gamble on complete discontinuities marks them as the models (whether acknowledged or not) for the Hustlers who would follow in their wake.

The Plutocrats were not homogeneous by any means. They included authentic engineering geniuses like Moore, hard-charging entrepreneurs like Hewlett and Packard, and consummate salespeople like the Watsons. They could be enthusiastic risk takers or blue-suited organization men. They did their best to lock in customers with proprietary solutions, but would gamble on entirely new models if they felt they could succeed. They built the business of computing, yet were never particularly committed to the dreams that animated the Patriarchs. During the Patriarchs' ascendency, however, the visionaries were out there, wondering why computers could not simulate different kinds of models. They had a kind of erotics of simulation in mind—the pleasure of doing *something* and then doing *everything* with these machines. The possibilities seemed endless. I call the next generation the Aquarians, and they were like painters eyeing a blank canvas or sculptors circling a block of marble.

The Aquarians: Douglas Engelbart and Alan Kay

In 20 or 30 years, you'll be able to hold in your hand as much computing knowledge as exists now in the whole city, or even the whole world.
—Douglas Engelbart

It's not the technology that lives. It's the dream that lives.
—Alan Kay

Doug Engelbart worked in the world that the Plutocrats ruled, but made the world in which we live. He drew inspiration from the visions of the Patriarchs, and they funded his campaign against the culture of the Plutocrats. As an army radio operator in the Philippines just after World War II, Engelbart read Bush's seminal article, "As We May Think," and its vision sustained him all the way through graduate school and into his first positions in California. Engelbart devoted his life to developing a new way to think about computers—not data processing machines as per the Plutocrats, but instead augmenters of the human intellect.[14] He did his major work of this period at the Stanford Research Institute (SRI), near Stanford University, where he was supported by funding from Licklider's DARPA, because Licklider understood that Engelbart's work took the modes of participation and linked them to the goal of augmentation.

Engelbart's vision harnessed computers to augment human intelligence in order to tackle the biggest questions that humans face: what were later to be called "wicked problems," or those that are socially and environmentally complex, ranging from eradicating poverty to planning cities to determining what constitutes a good life well lived. Engelbart set himself the grandest simulation objective of all: not simply simulating a single mind, but instead trying to simulate the best of group thinking and action, leading to the twined memes of symbiotic participation.

Of course, many people, both within and outside of computer science, have been concerned with wicked problems, but few of them ever had the kind of immediate, public impact that Engelbart did in 1968. For that was the year that he gave the "mother of all demos," a public display of his innovations and vision to an audience of his peers along with a younger generation that he would inspire. At SRI, Engelbart had developed a system featuring scaling windows, graphical user interfaces, live video teleconferencing, and hypertext.

A new input device of his invention—an odd-looking thing that could control elements anywhere on the screen—directed all of these windows and operations. Engelbart's patent application referred to it as an "X-Y position indicator for a display system." It featured a wooden shell small enough to be cupped in one hand, which covered two metal wheels, and the whole assemblage was connected to the computer by a long, thin cord that looked a bit like a tail. In honor of the tail, Engelbart and his team called it a "mouse."

This was the system that he set up for a live demonstration in 1968, teleconferencing from San Francisco to Palo Alto, showing the text-processing and electronic mail functions in separate, scalable windows, and doing all of this in real time, with a huge projector borrowed from the U.S. Department of Defense. In a world of computers dominated by alphanumeric input and output, command-line interfaces, and the barest minimum of give-and-take between users and machines, Engelbart's demo was a revelation, a melding of simulation and participation. Remember that all of this interfacing was taking place over a distance of thirty miles. If the SRI demo was not the full-fledged dawning of the Age of Aquarian computing, it came as close as computer science and engineering were ever going to come.

In the 1960s people were trying to group a vaguely hopeful set of feelings and instincts—that people could come together peacefully to make the world a better place, and that technologies from communication to drugs to space travel were combining to produce a new era—or what came to be known as the Age of Aquarius. The term Aquarian came to represent much that was hopeful, if not always practical, about the times.[15] Engelbart's demo, and the work of the other Aquarians in the 1960s and 1970s, may be little known, but it was central to the transformation of the "soul of the machine." They strove to humanize, decentralize, and personalize computers, and were opposed to virtually every aspect of the way that the Plutocrats had commodified and corporatized computing. What the Aquarians felt was missing in the Plutocratic era was the sense that humans had invented a new ally, not just for the battlefield, lab, or office, but in making a better, more creative life.

When people talk about Engelbart's presentation of the NLS ("oN-Line System") as the "mother of all demos" what they mean is that something about the reality of the thing—the real-time manipulation, the new input device, and the sheer totality of it all—changed the culture of computing right then and there, at least in the heads of those who could understand its implications. One of those best and brightest was the young Alan Kay. A polymath who had supported himself in grad school by playing jazz guitar, Kay had never felt comfortable in the confines of academia.[16] He had traveled down to the Bay Area from the University of Utah, where he was a grad student in the lab of Ivan Sutherland. Of all the people I group together as Aquarians, perhaps the least likely to accede to the label might be Sutherland. He was a straightlaced engineer, who after his greatest contributions to the field developed a profitable company working almost exclusively on military contracts.

Sutherland, buttoned-down though he may have been, was the figure who developed the visual side of computing, who opened up computing to the right brain world of artists and designers, and who crafted real-time responsive visualization technologies. These massive breakthroughs were so mind expanding that I feel forced to shoehorn him into this group. With a software called Sketchpad, which simulated as well as surpassed the visual capacities and productivity of paper and pen, Sutherland created the first robust and responsive computer graphics workstation. Sutherland was adamant that while he could simulate pencil on paper, his system opened up hitherto-impossible microdrawing, because the system allowed you to play with scale and zoom, and macrovisualization, because you could zoom out and create documents essentially without borders.

However the Provo, Utah–based Sutherland might have felt about the goings-on in the far more Aquarian San Francisco Bay area, for Kay it was like coming home. There he found a community that appreciated his approach to programming, especially his personal desire to develop computers that regular people and children, especially children, could play with. Engelbart's demo gave Kay and others like him proof on the ground that their ideas not only could work technically but also that they could excite a new kind of passion for computing. Within fifteen years, Kay had extended Engelbart's augmentation ideas into the realms of software—he pioneered object-oriented programming, refined the graphical user interface, and the Ethernet. Kay was a central figure in the development of Xerox PARC's Altair personal computer—a machine that the executives at corporate headquarters, three thousand miles east of Palo Alto, could not figure out how to market. The tension between the "suits" in the corporate suites on the East Coast and the techs in their experimental labs on the West Coast was a key conflict in the Aquarian period, and the legacy of those battles continues to animate the discourses of Silicon Valley to this day.

Even if Kay had not accomplished all of this, he would be remembered for creating the concept of "personal dynamic media." Kay and a team at PARC created a conceptual prototype they called the "Dynabook," and set a mark for all the laptops, personal computer tablets, and mobile computing devices to come. Kay foresaw the integration of digital modes of creativity into every aspect of human life from the earliest learning experiences to the most advanced scientific experiments.[17] He was like a number of other brilliant scientists involved in the Aquarian moment who looked to children to find inspiration, and drew from the pioneering work of Swiss developmental psychologist and pedagogical theorist Jean Piaget.[18] These computer scientists created what I have come to call a "kinderkult." Many scientists feel cut off from the rest of humanity, which for lack of training or aptitude does not master advanced mathematical reasoning. The Aquarians saw in the computer a way to bridge this gap, especially in regard to children. In addition, people like Kay felt that the computer would offer both themselves and children a way to tap the unfettered creativity we see in the young. This emphasis on pedagogy and play stimulated the Aquarians to advocate for the widest possible penetration of these technologies into society, pushing the meme of participation even further.

The Aquarian generation, like the Patriarch one before it, has been for the most part forgotten. That is not to say that historians, hardcore hackers, and the occasional technologist or techoartist have not drawn inspiration from their example, but in terms of general recognition, they do not rate nearly as high in the pantheon of cultural heroes as they should. The reason for this is painfully simple: they did not capitalize on their genius. Engelbart invented the mouse, and SRI sold the license for it for forty thousand dollars. Kay was part of the team that created the Alto, the first marketable personal computer, but Xerox could never quite figure out how to sell it.[19] Others, like Ted Nelson, the most explicitly Aquarian of them all, have been taken as cautionary tales by those who followed. The technology was there, the dream of participation was not just alive, it was thriving, but the Aquarians couldn't sell it to the masses. And selling to the masses is one way to be remembered, at least in the United States. Selling to the masses is what Hustlers were born to do.

The Hustlers: Steve Jobs and Bill Gates

Who can afford to do professional work for nothing?
—Bill Gates, 1976

Real artists ship.
—Steve Jobs, 1983

If J.C.R. Licklider and Douglas Engelbart are obscure talismans even among the best-informed computer users, Steve Jobs and Bill Gates are iconic, triumphal nerds.[20] Jobs at his height was perhaps the most intriguing businessperson in the world, and one of the few people to have built multibillion-dollar companies in two different industries, with Apple computers and Pixar animated films. Gates was the CEO of Microsoft, and as cab drivers from Calcutta to Cancún can tell you, for years the richest person in the world. More than anyone else, Jobs and Gates were responsible for making the computer part of people's daily lives, personalizing the technologies, and changing the ways in which information functions as a driving force in contemporary commerce as well as culture.

Jobs and Gates started out when personal computing, that idea advanced by Licklider the Patriarch and Kay the Aquarian, was the province of a tiny group of obsessed hobbyists. It was a business, but one with a smaller market than fly-fishing. As teenagers in the 1970s, Jobs and Gates were part of this small group of hobbyists who purchased kits to make simple, programmable computers to use (and play with) at home. Jobs, along with Steve Wozniak, were members of the best-known group of these enthusiasts, the Homebrew Computer Club of Cupertino, California. Gates, who had been programming since he found himself able to get access to a DEC mainframe in high school, was already writing software professionally while he was a student at Harvard. Jobs and Gates, along with their collaborators and competitors in the mid-1970s, were positioned at a fulcrum point, when a diversion turned into a business. What made them both rich and powerful was their ability to meld the attributes of the two generations that preceded them—fusing the hardheaded business logic of the Plutocrats with the visionary futurity of the Aquarians.

Jobs and Gates have an interesting competitive history, leapfrogging each other in the quest to achieve "insane greatness," in Jobs's words, and global market preeminence, for Gates.[21] Jobs and his partner, Wozniak, were the first to make the leap from hobbyists to industrialists with their Apple computers, launched in 1976. It was the Apple II that really broke loose, in 1977, attracting a huge user base, and establishing Jobs and Wozniak as the first publicly lauded millionaire whiz kids of Silicon Valley. As important as their early success with the Apple II was, however, their greatest impact came seven years later, when they took the inspiration of people like Engelbart and Kay, and created a mass-market personal computer that set a new standard for participation.

Before we get to that, we need to return to 1976, and move from Silicon Valley to New Mexico, where Gates and his partners, including former Harvard friends Paul Allen and Steve Ballmer, were writing programs for the Altair computer. That was the year that Gates wrote a famous letter to the hobbyist community complaining about rampant software theft. He claimed that nothing would please him more than establishing a business model for the community that would allow him to hire ten programmers to write software full time for the hobbyist market.

Just four years later, Microsoft was profitably writing closed-source code for a variety of platforms, and Gates was approached by IBM, then the biggest computer company in the world, to supply a stable operating system for Big Blue's new line of personal computers. It was at this point that IBM and its intellectual property lawyers entered into what is now universally seen to have been the worst contract in the history of global business. Gates, the son of a renowned intellectual property lawyer, negotiated a contract in which Microsoft licensed the operating system to IBM rather than selling it outright.

Microsoft was also free to license MS-DOS and related application software to any other company that produced computers. There was an explosion of personal computer clones piggybacking on IBM's success, but Microsoft, with no hardware manufacturing costs and a lock on the dominant operating system, turned out to be the big winner. Within a decade Microsoft was worth more than IBM, and Gates was the richest person in the world. Gates defeated the Plutocrats at their own game and ascended into their ranks. Even adjusted for inflation, the Watsons were never as rich as Gates became.

Gates has been admired for his strategy and condemned for his ruthlessness, but much like the Plutocrats before him, he was never defined by vision. Steve Jobs, the cofounder and CEO of Apple, on the other hand, has been known to create a "reality distortion effect" around himself because of the intensity of his vision for computing. He worked for early electronic games pioneer Atari in the late 1970s and visited Xerox PARC, where he saw the work infused with Engelbart and Kay's Aquarian vision. This spirit resonated with Jobs, who at one point had taken a personal pilgrimage to India and lived in an ashram. But even more so, the meme of participation entered his head on those visits to PARC.

The Apple II, released in 1977, was unique in having a graphics capability and a soundboard built in. Here was the first major computer for the masses, designed from the start as a multimedia machine. These Apple IIs became the de facto machines in classrooms around the country, and without a doubt prepared a generation of computer users for what was to come. Jobs understood that the graphical user interface would open up a whole new range of applications to nonexpert users, but even more would expand that user community exponentially. The Macintosh, released in 1984, brought the next level of integration of sound, image, and motion into personal computing.

It had built-in hypertext capabilities, the scaling windows of the Engelbart's famed demo, and of course, it made the mouse a ubiquitous device.

There is no doubting that Jobs traced his heritage to the Aquarians, and that in besting IBM at its own game, Gates transcended the Watsons. But the development and release of Microsoft Windows, to replace the command line DOS, was an interesting case of one generation simulating itself. U.S. courts ruled long ago that there was no copyright infringement of Windows on the Mac operating system, and contrary to myth, there was never a memo where Gates put on paper "Make it like the Mac," yet legalisms aside, Apple was more of an innovator, and Microsoft was more of an acquirer. For all their differences, though, the Hustlers will blend together in history's assessment, and the differences between Jobs and Gates will seem as slim to people in fifty years as the distinctions between John D. Rockefeller and Andrew Carnegie seem to us today.

What the Hustlers were able to do, regardless of who did what first and with what motives, was to apply the marketing savvy and managerial brilliance of the Plutocrats to disseminate to a huge, worldwide population the innovative spirit and inventive genius of the Aquarians. Over the next three decades, by hook or by crook, through mediocre upgrades and teeth-gnashing convergence headaches, more and more people found themselves in possessions of machines that had ever-greater computing power, which translated into an ever-increasing ability to simulate the other media devices that they already possessed, but here integrated into an ever more seamless (at least in theory) system. That was all well and good, but even amid this transformation, one of the key memes of the Patriarchs and Aquarians was missing. What good was all this computing power and creative potential if people could not share it with each other, if they could not simulate that wondrous nineteenth-century information appliance, the telephone? In other words, who would turn this installed base of atomized personal computers into a communication system? Who would link them together to share their entertainments and experiences, and even pool resources to tackle those wicked problems? The Hosts, of course.

The Hosts: Tim Berners-Lee and Linus Torvalds

You affect the world by what you browse.
—Tim Berners-Lee

Software is like sex: it's better when it's free.
—Linus Torvalds

After the triumph of Apple, the Macintosh, Pixar, iTunes, the iPhone and the iPad, it is hard to remember that Jobs also had his share of failures. Buried in his corporate biography is NeXT Computer, the short-lived company that he founded after leaving Apple in the mid-1980s. The beautifully designed NeXT Cube was a commercial disaster. Like the now-forgotten Apple Lisa, the predecessor to the Macintosh, it was too expensive to develop more than a tiny niche market among researchers and academics. But the NeXT Cube was to mark a transformation in the way that people and the market interacted not with "personal" but rather with "interpersonal computers." From the beginning, Jobs and his team designed NeXT Cubes to communicate with other NeXT Cubes. Here was a machine that networked in ways that the Patriarchs had foretold decades before, and that Engelbart and Kay had spent the intervening years refining, with object-oriented programming, shared work environments, and integrated visualization and audio tools. In other words, this was a machine that was doing much of what people expected of their machines after the year 2000, but fifteen years earlier.

In the late 1980s, in a physics laboratory in Geneva, a British scientist worked on a NeXT Cube, testing its capacity to deal with one of the main issues that he confronted in his field. Physics, like its subject, frequently seems to move at the speed of light, but the mechanism for publishing research, via paper journals, is often maddeningly slow. In an effort to speed up the transfer of information both within and outside his lab, Berners-Lee started thinking about electronic publishing. He had experience working with electronic typesetting systems at an earlier job and had been thinking about hypertext systems for more than a decade. In 1989, he wrote a paper about combining a visual hypertextual system with the communication protocols of the Internet to link these documents together over the network.[22] Berners-Lee pursued a double goal of simulation: he wanted hypertext to simulate the communicativity of networks, and he wanted networks to simulate the visualized linkages of hypertext. Two years later, in 1991, on a NeXTStep server, he built just such a system, and uploaded the first page on to what he called the World Wide Web: <http://info.cern.ch/hypertext/WWW/TheProject.html>. Here was Licklider the Patriarch's Intergalactic Network melded to the Aquarian Nelson's unrealized universal publishing database Xanadu.

It is worth the effort to decode what <http://info.cern.ch>, a cryptic set of letters, means. First, it is called a Universal Resource Locator (better known as a URL), and it is a linguistic pointer to a set of numbers that it is assigned to—namely, 0.001.0012.2275. Second, "http" stands for HyperText Transfer Protocol, which ties together the linking functions of hypertexts with the communication functions of the Internet. "Info" is the grail here, as the page contains something that one person or group has uploaded, and that another person or group will download. Next, "cern" is the laboratory where it was created, which at that time had a greater density of Internet connections than any other locale in Europe, and "ch" is an abbreviation for the name *Confoederatio Helvetica*, which is the Latin name for Switzerland.[23] This first page told people how to set up their own page and then link to other people, and for a time was also the first directory to the Web, as Berners-Lee would list the new pages as people sent him links to them.

The effect of this was viral and global. There had been ways of distributing and searching documents via the Internet before, like now-obscure protocols Telnet, Gopher, and Fetch, but Berners-Lee had created the first intuitive, graphical user interface for the Internet. And he was adamant from the start that producing material for the Web should be as easy as accessing it. In this, he was a powerful advocate for transparency, cooperation, and productive interaction. Like the Aquarians, he saw this work as a gift, and he offered all of this free of royalties for his inventions and patents—meaning that anybody or everybody could develop the Web. A few years later, Marc Andreesen and a group of other Web developers launched their own browser, known as Netscape, kick-starting the Web bubble of the mid- to late 1990s. This ignited a frenzy of wealth building, earned Berners-Lee a knighthood, and encouraged otherwise-sane people to claim that the invention of the Web had more potential than the discovery of fire. Hyperbole, to be sure, but there is no doubt that the Web was transforming the culture machine.

The 1999 Golden Nica, the most prestigious award in the media arts, generated controversy when the committee decided to give it jointly to a programmer and the operating system he developed. How and why the jurors at Austria's Ars Electronica Festival decided to so honor the Finnish computer scientist Torvalds, a person who had never claimed to be an artist, and Linux, a collection of code rather than an artwork, will be something that we will return to at the end of this section. The jury's statement acknowledged the incongruity, yet argued that Torvalds, his software, and the community that created it "birthed an aesthetic showing how something can be built on the Net through an intentional, but not necessarily direct, description."[24]

To understand why Torvalds is so important to the meme of participation, we have to revisit the Plutocrats. In the first twenty years or so of computing after World War II, when institutions bought mainframes from vendors like IBM and DEC, the software on these machines came with complete documentation of the source code. There were far fewer applications in those days, and the "users" were almost by definition experts and programmers themselves, so it made sense to let these people develop their own software for their systems. In the late 1960s and earlier 1970s, however, systems and softwares were debundled, which meant that people were starting to treat the softwares as commodities of intrinsic value.

Eventually this proprietary software was seen by most people as the only kind of software there was: a commodity with restrictions against redistribution, an opaque entity that did not make its source code available, and a stable tool that users were not to modify. This shift made sense to the commercial software vendors, and without question created the (relatively) stable set of platforms and softwares that encouraged small businesses as well as individuals to follow the lead of large institutions in digitizing their work and lives. And of course, fortunes were made and lost.

Yet throughout the 1970s and 1980s, a few key figures, such as Richard Stallman in particular, felt that the earlier ethos of open-source software was something worth building on. Stallman wanted to extend the sense of scientific collegiality and openness that he valued in the hacker community. So he developed a new operating system compatible with the UNIX family, but distinct from it, and made it available to other like-minded programmers. In hackerish fashion, he christened it GNU (an acronym for the recursive phrase "Gnu's Not Unix"), and in 1985, he started the Free Software Foundation to support his work and that of his growing community of collaborators. In the "GNU Manifesto," Stallman defined the ethos of free, or open, software:

1. You have the freedom to run the program, for any purpose.

2. You have freedom to modify the program to suit your needs. (To make this freedom effective in practice, you must have access to the source code, since making changes in a program without having the source code is exceedingly difficult.)

3. You have the freedom to redistribute copies, either gratis or for a fee.

4. You have the freedom to distribute modified versions of the program, so that the community can benefit from your improvements.[25]

The community of developers and users for GNU was committed, vocal, and productive, but relatively small, restricted to academic labs, a few dedicated professional programmers working off-hours, and hackers interested in free software. In 1991, though, a young grad student in Finland radically altered the landscape for collaborative free software development.[26] Torvalds developed a new operating system that he came to call Linux. His great innovation was to realize that the Internet would change everything about the open-source/free-software community. Torvalds released Linux in a way that ensured that the thousands and then tens of thousands of software designers, hackers, hobbyists, and assorted other users/makers could modify, improve, and add to the central Linux operating system. At one point, a study was done demonstrating that over one billion dollars worth of programming hours had been devoted to Linux, at no or minimal cost.[27]

There were reasons that the apparent gift economy of Linux advanced as it did. For one thing, Torvalds and the core group of early enthusiasts maintained a tight ship. For another, because there were so many people pushing and prodding at Linux—far more than could ever see the closed-source commercial code from companies like Microsoft, HP, Apple, Sun, and Oracle—Linux became a stable, bug-free operating system. This made it particularly good for people and groups running servers. Everyone from renegade hackers, to corporations that needed the utmost in stability, to governments like Brazil and China that did not want to become addicted to proprietary Microsoft softwares came to embrace Linux.

It is important to see what expanded networks bring to open-source development. The programmer Eric Raymond describes closed-source software as the information age equivalent of the cathedral, a vast structure driven by the church and architects, even if it took generations to build. He contrasts this with open source, which he sees as a bazaar, with tents going up, wares being laid out, reputations rising and falling and the individual merchants constantly wrangling with the customers and each other.[28] The cathedral is a vast and imposing structure, but if something goes wrong, it is both difficult and expensive to fix (and in any case, there may be no one looking at the right place in the wall for structural flaws). With Linux, however, all faults are held up for immediate inspection and potential fixes. Raymond coined the phrase "Given enough eyeballs, all bugs are shallow." He came to dub this Linus's law.

There is a tendency in the open-source movement to elide the differences between developing software and making other forms of culture. The software model is based in large measure on the oscillation between problems and solutions, bugs and hacks. This can be a good way to harden software and add functionality, but when it comes to art, music, or literature, there are different modes of production, judgment, and success. Not every sphere of production is improved by collaboration, nor can every aesthetic issue be "solved" by the application of more eyeballs (or earlobes). The model of cooperative, open-source cultural production, as represented by the Creative Commons movement, has its roots in free software, although it differs in ways both large and small. For our purposes, though, it is the Creative Commons model that is at this point the most meaningful contributor to the forces of uploading, as discussed earlier.

Another caveat: free software does not necessarily mean "cost free." Neither Stallman nor Torvalds forbade selling software created with their kernels, and companies like Red Hat made a tidy sum packaging and supporting the Linux operating system. What makes software free is the access to the source code. The decision to market "free" software is left to the developer(s) involved. Over the past few years, the GNU/Linux open source expanded and became a robust—if still technologically daunting—alternative to the copyrighted, proprietary software of behemoths like Microsoft and Oracle, while the model of collaborative, cooperative development yielded results that could go toe-to-toe with established, for-profit business models.

The Searchers: Larry Page, Sergey Brin, and Others

Don't Be Evil.
—Google corporate motto

Any sufficiently advanced technology is indistinguishable from magic.
—Arthur C. Clarke

Half a century into the computerization of culture, whatever linear narratives of origin we have been able to map out here definitively break down. The bursting of the dot-com bubble was reminiscent of the disastrous fate of railroad companies in the United States in the late nineteenth century. When they went bankrupt after their own technological fever dream, they left two legacies—one short-lived, and the other vastly more important. The first legacy involved the defrauded and angry shareholders. They are, for all but a few financial histories, the forgotten victims of bad timing in the business cycle. The other legacy of that stock bubble was a continent connected by rail, with excess capacity along with the means to build the world's largest unified network of producers and consumers. The original entrepreneurs and investors in railroads may have lost millions, but the United States as a whole benefited immeasurably from their efforts. The Web 1.0 bubble laid much "dark fiber" across the world, as companies built far broader networks than they could ever use profitably, and after the crash, others have since benefited from that infrastructure to restructure the ways we conceive of and engage with the Internet.

No one company has so palpably benefited and defined this shift than Google, the search algorithm that became a company and then a verb, as noted earlier. Google was an intentional misspelling of the word "googol," the mathematical term for a one followed by ten zeros. The company became a networked Ourobors, that creature from Greek mythology that devours its own tail and encircles the world. What cofounders Larry Page and Sergey Brin created was a relentless innovation and acquisition machine, powered by users and advertisers alike. They challenged the old masters, from Microsoft to Yahoo! and infiltrated everything from libraries to desktops, enmeshing everyone from pornographers to cartographers, from anticorporate bloggers to CEOs. Yet by the time they accomplished all of this, they had blurred the lines between simulation and participation, and made people wonder about the "goodness" inherent in their corporate motto forsaking evil.

To understand Google, it is imperative to understand that the company's success was a dividend of the groundwork laid by the Hosts. The first months after Berners-Lee launched the World Wide Web in 1991, there were only a few hundred pages, and a few months after Netscape was launched in 1994, a few million. That same year, two Stanford students started uploading and updating their favorite sites as "Jerry and David's Guide to the World Wide Web." The two students, Jerry Yang and David Filo, went on to found Yahoo! They continued their commercial venture along a "curatorial" strategy, organizing different sites and pages under larger categories. The ever-denser layout of the Yahoo! home page made sense under this system, creating categories and subcategories for users to orient themselves. Yet when Page and Brin showed up at Stanford a few years later, the Web moved from millions to hundred of millions to billions and then hundreds of billions of pages, and the significance of searching as well as the so-called engines to perform these searches became the defining need of the era. What Page identified was that it was not simply important to find something; you really wanted to know what the importance was of what you had found. There were few reliable ways to use machine intelligence to create usable hierarchies, and you could not hire enough people to sift through something as densely complex as the Web by 2000. What Page decided to do was to let the Web itself "rank" the pages within it, creating a search algorithm that used the number of links to and from a page to determine its "utility" to users.[29] With its relentless concentration on user experience and convenience, and an ever-mutating and sophisticated approach to revenue generation, Google exploded over its first decade.

As with so many of the histories we have been exploring, there were myths of origin. Their first angel investor, a cofounder of Sun Microsystems, was in a rush, and after looking at Page and Brin's demo, said, "Instead of us discussing all the details, why don't I just write you a check?" His check for hundred thousand dollars was made out to Google, Inc., a legal entity that did not as yet exist, and it took Page and Brin a few months until they were able to cash it (having raised close to a million more in the interim). The company's insistence on participation is obvious, but Google is also one of the great sources of simulation in the Web n.0 era. From the commercial simulations mimicking desktop applications in order to move them into browsers (as Gmail copied the best features of commercial email programs and made them available on the Web), to the metasimulationist efforts of Google Maps to chart the whole world both cartographically and photographically, Google has embraced simulation as an attractor to its core search and advertising model.

The company made a series of conscious choices about the design of its site, business model, and image. For years, Google's home page featured a design degree-zero approach: a text box and a button, with the logo shifting only with holiday treatments and the occasional April Fool's Day joke. This conscious decision to limit the information to the back end was a remarkable contrast to the Yahoo! home page plethora of categories, subcategories, and information feeds. The decline of Yahoo!'s approach indicates the success of Google's.[30] The latter's corporate culture is marked by holdovers from the dot-com era: almost no one wears a suit, and the cofounders are routinely quoted as saying that they are only serious about search; the Googleplex, the company's Mountain View corporate office, features the usual foosball tables and snack rooms, but adds in lava lamps and exercise balls; and in an added northern California touch, the corporate cafeteria's chef used to cook for the Grateful Dead. Yet Google's relentless focus on revenue generation has proven far more successful than the amorphous "brand building" of the Web 1.0 flameouts.

Google, irrespective of its professed motto to not be evil, cannot help but generate questions and opposition, simply due to its size, ubiquity, and profitability. The "free" use of Google depends on advertising. As Google has bought up both Web 2.0 content generators, like Blogger for blogging software along with hosting space and YouTube for video, and the major advertising enablers, like Doubleclick and Adstar, it becomes a stakeholder on not just two sides but instead apparently every side of the relationships between users, advertisers, and content providers (be they professional, amateur, or that ever-growing group positioned either in between these two terms or apart from them completely). For all of its engagement with the participatory qualities of Web 2.0, Google itself is something of a black box. The exact composition of this algorithm and how it is combined with other factors is the core intellectual property of Google, and thus its most zealously guarded secret. No one even knows how many servers it has, nor how these servers are distributed globally. While Google searches are free and the results are ubiquitous, the company itself is the antithesis of the open-source movement. As Google becomes more and more enmeshed in our lives, the issues of bias, censorship, and above all commercial interests come to the fore.

The closer histories come to the present moment, the greater the danger that their authors will mistake passing fancies for lasting impacts and ignore the small but influential in favor of the bombastic though ultimately meaningless. It seems unlikely that the future will look back at the desire to aggregate information, people, and media via networks as insignificant. By the turn of the millennium, the memes of simulation and participation had become so intertwined as well as spread so widely on networks globally that the next step had to be its organization and access. The generational narratives of the first half century of the culture machine were breaking down at the very instance that participation and simulation became ubiquitous and the very definition of the unimodern moment.

NOTES

INTRODUCTION: THREE SIBLINGS

1. Marie Winn, *The Plug-in Drug: Television, Children, and the Family* (New York: Viking, 1977).
2. In 1950, fewer than 10 percent of families in the United States owned televisions; within four years, the percentage rose to more than half. Three years later, it was over 75 percent, and by 1964, more than 90 percent of U.S. families had a television—almost total penetration in less than half a generation's time.
3. I cannot go as far or as ironically as journalist Lee Siegel does in an essay: "Oprah Winfrey is to television what Bach is to music, Giotto to painting, Joyce to literature." Lee Siegel, "Thank You for Sharing: The Strange Genius of Oprah," *New Republic*, June 5, 2006, available at <http://www.tnr.com/politics/story.html?id=15d21968-03ba-437e-a5fd-f2712b592b21>.
4. On the Oprahization of discourse and rise of the first-person narrative by scholars in response to concerns about the viability of overarching cultural theory, see Cynthia G. Franklin, *Academic Lives: Memoir, Cultural Theory, and the University Today* (Athens: University of Georgia Press, 2009).
5. André Glucksmann, "From the H-bomb to the Human Bomb," *City Journal* 17, no. 4 (Fall 2007), 56–63.

CHAPTER 1: THE SECRET WAR

1. One cause of confusion is that given the importance of packet sharing to transfers of any kind over the Internet, a single file will be broken down, and then uploaded and downloaded many times by many servers in the course of its "travels," regardless of where it originates and where it is going.
2. In 2006, the numbers ran 1 percent uploading, 10 percent commenting and modifying, with the rest just surfing through. Charles Arthur, "What Is the 1% Rule?" *Guardian Unlimited*, July 20, 2006, available at <http://technology.guardian.co.uk/weekly/story/0,,1823959,00.html>. In 2009, the Harvard

Business School's Mikolaj Jan Piskorski found that with Twitter, 90 percent of the tweets were created by 10 percent of the users. According to the same researcher, "Seventy percent of all actions" on social media sites like Facebook "are related to viewing pictures or viewing other people's profiles"—in other words, downloading rather than uploading. Sean Silverthorne, "Understanding Users of Social Networks," *Harvard Business School Working Knowledge*, September 14, 2009, available at <http://hbswk.hbs.edu/item/6156.html>.

3. DVRs like TiVo can be seen as time shifters of this downloading rather than a qualitatively different experience. Quantitatively, DVR users watch more television after they purchase the device than they do before owning one.
4. In the West, it tends to be a disease of the poor, as the rich have adopted lower-calorie diets for reasons of aesthetics as much as health. In the developing world, diabetes is becoming a disease of the rich, who can afford to consume more food. Charles F. Burant, ed., *Medical Management of Type 2 Diabetes* (Alexandria, VA: American Diabetes Association, 2004).
5. According to Harvard Medical School's Joslin Diabetes Center; see <http://www.joslin.org/>.
6. One corporation in particular, the Archer Daniels Midland Company, made HFCS a huge part of its growth plan in the 1970s and 1980s. This story is now best known through its telling in Michael Pollan, *The Omnivore's Dilemma: A Natural History of Four Meals* (New York: Penguin Press, 2006).
7. Carl Honore, *In Praise of Slowness: How a Worldwide Movement Is Challenging the Cult of Speed* (New York: HarperOne, 2004). For Slow Architecture, see <http://www.archinect.com/news/article.php?id=P2852_0_24_0>. There are also resources at <http://www.slowdesign.com>. On these issues in graphic design, see Michael Beirut, "In Praise of Slow Design," in *Seventy-nine Short Essays on Design* (New York: Princeton Architectural Press, 2007), 237–240; originally published in blog form, available at <http://observatory.designobserver.com/entry.html?entry=3947>.
8. Many television scholars have trouble even going this far. An essay on the show *Lost* ties itself into preemptive knots to ward off critiques of snobbery and canon formation for asserting that *Lost* was "a great show," and better made, better acted, and worthy of deeper consideration than other shows on television. The author plaintively insists to his peers that "even within the realm of the vulgar and base, we must acknowledge that some crap is better than other crap." Jason Mittell, "Lost in a Great Story," in *Lost: Perspectives on a Hit Television Show*, ed. Roberta Pearson (London: I. B. Tauris, 2009); also available at <http://justtv.wordpress.com/2007/10/23/lost-in-a-great-story/>.
9. See Barbara Klinger, *Beyond the Multiplex: Cinema, New Technologies, and the Home* (Berkeley: University of California Press, 2006).
10. Todd Gitlin uses the term torrent to define the entirety of media inundation, including both television and digital media. Todd Gitlin, *Media Unlimited:*

How the Torrent of Images and Sounds Overwhelms Our Lives (New York: Metropolitan Books, 2002).

11. From the abomination of ninth graders forced to watch *Channel One* in class to the overwhelming choice that those who go off to college make a few years later to install personal televisions in their dorm rooms. See Mark Crispin Miller's white paper prepared for Fairness and Accuracy in Media in 1997, "How to Be Stupid: The Teachings of Channel One," available at <http://www.fair.org/index.php?page=1384>.
12. This frequently repeated metaphor appears to be entirely unsupported by evidence—frogs tend to hop around in any case, especially when their environment changes for the worse—but it is such a useful image that I refuse to abandon it simply because it is untrue.
13. This quote comes from a 1978 essay by Philip K. Dick, "How to Build a Universe That Doesn't Fall Apart Two Days Later," which is included in Mark Hurst and Paul Williams, eds., *I Hope I Shall Arrive Soon* (New York: Doubleday, 1985), available at <http://bigpicture.typepad.com/writing/scifi/index.html>.
14. Mickey Alam Khan, "Potato Spurs Interest in Broadband TV," *DMNews* (May 18, 2006), available at <http://www.dmnews.com/Potato-Spurs-Interest-in-Broadband-TV/article/91210/>.
15. See Philip K. Dick, *Do Androids Dream of Electric Sheep?* (New York: New American Library, 1968), the novel upon which Ridley Scott based his film *Blade Runner* (1982).
16. Reyner Banham, *Los Angeles: The Architecture of Four Ecologies*, intro. Anthony Vidler (1971; repr., Berkeley: University of California Press, 2001). [1971]). The full video is available at <http://video.google.com/videoplay?docid=1524953392810656786>.
17. I first saw the *Livre de prières* at the Hammer Museum in Los Angeles in 2002 at a show of books and objects from library special collections in Southern California. For documentation of the show and the *Livre de prières*, see Cynthia Burlingham and Bruce Whiteman, eds., *The World from Here: Treasures of the Great Libraries of Los Angeles* (Los Angeles: Getty Trust Publications, 2002), 62–63.
18. Estimates have ranged between one and five hundred thousand cards to program the loom to produce the *Livre de prières*, but historians do not know the exact number.

CHAPTER 2: STICKY

1. Two central figures in contemporary capitulationism are Henry Jenkins and Steven Johnson. See Henry Jenkins, *Convergence Culture: Where Old and New Media Collide* (New York: New York University Press, 2006); Steven Johnson, *Everything Bad Is Good for You: How Today's Popular Culture Is Actually*

Making Us Smarter (New York: Riverhead Books, 2005). The work of *New Yorker* phenomenon Malcolm Gladwell is a veritable trifecta of capitulationist journalism: *Tipping Point: How Little Things Can Make a Big Difference* (New York: Little, Brown and Company, 2000); *Blink: The Power of Thinking without Thinking* (Little, Brown and Company, 2005); *Outliers: The Story of Success* (New York: Little, Brown and Company, 2008).

2. Matthew Arnold, *Essays in Criticism* (1865; repr., London: Macmillan and Company, 1895), 37.
3. A case can be made from the computer science perspective that "emulation" works better as a description of this process than "simulation," but the more general connotations of the second word lend it to my use here.
4. James Jerome Gibson, "The Theory of Affordances," *Perceiving, Acting, and Knowing: Toward an Ecological Psychology*, ed. Robert Shaw and John Bransford (Hillsdale, NJ: Erlbaum, 1977); Donald Norman, *The Design of Everyday Things* (New York: Doubleday, 1988).
5. See Christopher Strachey, "The 'Thinking' Machine," *Encounter* 3, no. 4 (1954): 25–31. For an overview, see Noah Wardrip-Fruin, "Reading Digital Literature: Surface, Data, Interaction, and Expressive Processing," in *A Companion to Digital Literary Studies*, ed. Susan Schreibman and Ray Siemens (Oxford: Blackwell, 2008), available at <http://www.digitalhumanities.org/companionDLS/>.
6. Ellen J. Langer defines mindfullness as follows: openness to novelty; alertness to distinction; sensitivity to different contexts; implicit, if not explicit, awareness of multiple perspectives; and orientation in the present. Adapted from Ellen J. Langer, *The Power of Mindful Learning* (Reading, MA: Addison-Wesley, 1997). See also Ellen J. Langer and Mihnea Moldoveanu, "The Construct of Mindfulness," *Journal of Social Issues* 56, no. 1 (2000): 1–9, available at <http://cms.dartmouth.edu/conferences/langer1.pdf>.
7. Kenneth V. Iserson and John C. Moskop, "Triage in Medicine, Part I: Concept, History, and Types," *Annals of Emergency Medicine* 49, no. 3 (March 2007): 275–281.
8. The tech journalist Danny O'Brien coined the term in 2004; see <http://lifehacker.com/software/interviews/interview-father-of-life-hacks-danny-obrien-036370.php>.
9. See <http://lifehacker.com/028869/lifehacker-frequently-asked-questions>.
10. Developer Fred Stutzman, a teaching fellow at the University of North Carolina's School of Information and Library Science, quoted in Steve Kolowich, "Computer Program Wants to Free Scholars from Computer Distractions," *Chronicle of Higher Education*, February 5, 2009, available at <http://chronicle.com/wiredcampus/article/3597/computer-program-aims-to-free-scholars-from-computer-distractions>.
11. See Lisa A. Petrides, Sara I. McClelland, and Thad R. Nodine, "Costs and Benefits of the Work-around: Inventive Solution or Costly Alternative," *International*

Journal of Educational Management 8, no. 2 (2004): 100–108. I was introduced to this article at a workshop titled "The Work-around as a Social Relation" sponsored by the University of California Humanities Research Institute, the University of California at Irvine Department of Anthropology, and the Intel Corporations's People and Practices Research Group in 2008.

12. Flow, as defined three decades ago by the critic Raymond Williams, describes how broadcasters would arrange a night of television viewing to keep the viewer glued to a channel, unwilling to change to another. Williams's point was that contrary to what viewers thought, they were not spectators watching shows but were instead eyeballs being sold to advertisers. Raymond Williams, *Television: Technology and Cultural Form* (New York: Schocken Books, 1975).
13. Fredric Wertham, *Seduction of the Innocent* (1953; repr., Laurel, NY: Main Road Books, 2001); Jerry Mander, *Four Arguments for the Elimination of Television* (New York: Harper Perennial, 1978); Andrew Keen, *The Cult of the Amateur: How Today's Internet Is Killing Our Culture* (New York: Doubleday, 2007); Lee Siegel, *Against the Machine: Being Human in the Age of the Electronic Mob* (New York: Spiegel and Grau, 2008); Maggie Jackson, *Distracted: The Erosion of Attention and the Coming Dark Age* (Amherst, NY: Prometheus Books, 2008).
14. Lord Kitchner, "Kitch's Bebop Calypso," track 5 on the Honest Jon's compilation CD titled *London Is the Place for Me: Trinidadian Calypso in London, 1950–1956* (2001).
15. During the first Internet boom of the 1990s, a Web site's "stickiness" came to refer to the ability to hold on to users and bring them back for more. This seemed, in the awful parlance of the era, a "no-brainer," in that whatever you could do to hold on to eyeballs to flash more advertisements at them was *de facto* a good idea. This led to a concept of the site as a casino, which is designed with numerous entrances, extensive distractions, and few exits. This kind of stickiness waned as users became exasperated when they wanted to do things fast on a site—like register, purchase, or download information—and they could not achieve their aims without a host of other things happening to them. In the end, sticky was good for some experiences and not so good for others. The realm of culture overlaps with that of commerce, but is not wholly subsumed by it. And in culture, stickiness is not simply about trapping users in a labyrinth, or encouraging the use of one's own product or site rather than another's (which is also called lock-in or switching costs). See "The Fall of the Cult of Stickiness," available at <http://www.clickz.com/experts/archives/ebiz/ecom_comm/article.php/839661>. Advertising also uses the term sticky to define its ability to hold advertisers. See Lawrence Lessig, *Free Culture: The Nature and Future of Creativity* (New York: Penguin, 2005), 127.
16. Chris Lehmann defines mass culture as "the culture of market prerogatives," identifies television as "still the signature medium for mass cultural expression," and sees mass culture as profoundly different from "popular culture." He talks

of the intense semantic shift between the two words and identifies the reigning ideology as "mass culture *is* culture; all culture is popular." Lehmann points to this as the reason that "we can no longer distinguish between a self-created cultural life and a merely manufactured one." Chris Lehmann, *Revolt of the Masscult* (Chicago: Prickly Paradigm, 2003), 2–3.

17. As Theodor Adorno noted, "Meaning is present even in the statement that there is no meaning." Theodor W. Adorno, *Aesthetic Theory*, trans. C. Lenhardt (1970; repr., London: Routledge, 1984), 154.
18. In Gregory Fried and Richard Polt's translation of Martin Heidegger, *An Introduction to Metaphysics* (New Haven, CT: Yale University Press, 2000), the first line of the first chapter, "The Fundamental Question of Metaphysics," reads: "Why are there beings at all instead of nothing?"
19. In an article titled "Why Has Critique Run Out of Steam?" the French pioneer of science studies Bruno Latour offered a possible explanation: "I have always fantasized that what took a great deal of effort, cost a lot of sweat and money for people like Nietzsche and Benjamin, can now be had for nothing, much like the supercomputers of the 1950s, which used to fill large hall and expend vast amounts of electricity but are now accessible for a dime and no bigger than a fingernail." Bruno Latour, *Critical Inquiry* 30, no. 2 (Winter 2004): 25–248, available at <http://criticalinquiry.uchicago.edu/issues/v30/30n2.Latour.html>.
20. See Henry Jenkins, *Convergence Culture: Where Old and New Media Collide* (New York: New York University Press, 2006); *Fans, Bloggers, and Gamers: Exploring Participatory Culture* (New York: New York University Press, 2006).
21. Kingsley Amis, *Lucky Jim* (1953; repr., London: Penguin, 1963), 247.
22. My discussion of power, play, and the importance of tweaking draws from Michael E. Hobart and Zachary S. Schiffman, *Information Ages: Literacy, Numeracy, and the Computer Revolution* (Baltimore: Johns Hopkins University Press, 1998).
23. We should note that within drug culture, specifically among crystal methamphetamine users, "tweaking" describes the ever more obsessive-compulsive and destructive behavior brought on by the drug—as opposed to the more beneficent "tripping" on marijuana and other hallucinogens.
24. This language is taken from the twitter.com home page (accessed April 28, 2008).
25. "Linda Stone's Thoughts on Attention and Specifically, Continuous Partial Attention," available at <http://continuouspartialattention.jot.com/WikiHome> (accessed August 26, 2007).
26. Quoted in Virginia Postrel, *The Substance of Style: How the Rise of Aesthetic Value Is Remaking Commerce, Culture, and Consciousness* (New York: HarperCollins, 2003), 11.

CHAPTER 3: UNIMODERNISM

1. Malcolm Gladwell, *The Tipping Point: How Little Things Can Make a Big Difference* (Boston: Little, Brown and Company, 2002).
2. Paul Rand, quoted in Michael Kroeger, *Paul Rand: Conversations with Students* (New York: Princeton Architectural Press), 28–29.
3. The classic work is Frances A. Yates, *The Art of Memory* (1966; repr., Chicago: University of Chicago Press, 1974).
4. In 1911, Einstein stated the paradox this way: "If we were to place a living organism in a box, one could arrange that the organism, after an arbitrarily lengthy flight, could be returned to its original spot in a scarcely altered condition, while corresponding organisms, which had remained in their original positions, had long since given way to new generations. So far as the moving organism was concerned the lengthy time of the journey was a mere instant, provided the motion took place with almost the speed of light." Quoted in Anthony J. G. Hey and Patrick Walters, *Einstein's Mirror* (Cambridge: Cambridge University Press, 1997), 64.
5. If Krikalev were an American, he would be an international media star, with boxes of Krikalev-emblazoned Wheaties, X-treme soda sponsorships, and guest slots on late-night television chat shows. But Krikalev has two major strikes against him: not only is he "foreign," he carries the stigma of the cold war—a one-two punch against his potential as an endorser.
6. Andrei Ujica's film *Out of the Present* (Germany/Russia, 1995) is a cinematic essay on Krikalev's experiences in 1991. See the dialogue between Krikalev and Ujica, translated by Sara Ogger, "Toward the End of Gravity," *Grey Room* 10 (Winter 2003): 46-57.
7. For a sense of the early wonder of word processing, see James Fallows, "Living with a Computer," *Atlantic* (July 1982), available at <http://www.theatlantic.com/doc/198207/fallows-computer>.
8. Tristan Tzara used this technique with the Dadaists decades earlier, but Gyson's collaborations with Burrough's spread the concept far more widely.
9. Heath Bunting, "Own, Be Owned, or Be Invisible," 1998, available at <http://www.irational.org/heath/_readme.html>.
10. See, for example, Trevor Merriden, *Irresistible Forces: The Business Legacy of Napster and the Growth of the Underground Internet* (Oxford: Capstone, 2001); Joseph Menn, *All the Rave: The Rise and Fall of Shawn Fanning's Napster* (New York: Crown Business, 2003); John Alderman, *Sonic Boom: Napster, MP3, and the New Pioneers of Music* (Cambridge, MA: Perseus, 2001).
11. Bruce Mau, *Life Style*, ed. Kyo Maclear with Bart Testa (London: Phaidon, 2000), 65.
12. PostScript, like so many other apparent 1980s' innovations like the Ethernet, personal computing, and even portable notebooks, had its origins in the

ill-fated Xerox PARC Star system—that infamous victim of geek versus suit strife. John Warnock worked on that project, and ended up leaving Xerox PARC with two others to cofound a spin-off called Adobe, and with encouragement from Steve Jobs at Apple, Warnock and his partners developed the language to drive laser printers.

13. I am indebted to years of discussions with Lev Manovich on these topics, and can refer to *Software Takes Command*, an e-book in progress, available at <http://lab.softwarestudies.com/2008/11/softbook.html>.
14. The Bruce Willis press conference at Cannes in 1997 is described in Rob Nelson, "Inspecting the Cannes: In Its 50th Year, Is the World's Most Prestigious Film Festival Showing Its Age?" *Minneapolis City Pages*, May 28, 1997, available at <http://www.citypages.com/1997-05-28/arts/inspecting-the-cannes>. Willis was specifically referring to his starring role in Luc Besson's *The Fifth Element* (1997), and was also quoted as saying that "the written word is going the way of the dinosaur, anyway."
15. For a relentlessly negative view of the effects of networked technologies on twenty-first-century culture, see Mark Bauerlein, *The Dumbest Generation: How the Digital Age Stupefies Young Americans and Jeopardizes Our Future (Or, Don't Trust Anyone under 30)* (New York: Tarcher, 2008).
16. The designer made it clear in response to the numerous contemporary critics of his now-lauded collection that historicism alone was not the point: "Fashion is the reflection of our time, and if it does not express the atmosphere of its time, it means nothing." Yves Saint Laurent interviewed by the *New York Times* in 1971, quoted in Alicia Drake, *The Beautiful Fall: Lagerfeld, Saint Laurent, and Glorious Excess in 1970s' Paris* (New York: Little, Brown and Company, 2006), 114.
17. See Victoria Vesna, ed., *Database Aesthetics: Art in the Age of Information Overflow* (Minneapolis: University of Minnesota Press, 2007).
18. At the Metropolitan Opera in New York City, only *Aida* has had more productions mounted over the last hundred years.
19. On just how enduring the myth of bohemia is, see Richard Lloyd, *Neo-Bohemia: Art and Commerce in the Post-Industrial City* (London: Routledge, 2005). For a critical appraisal of the film, see Marsha Kinder, "Moulin Rouge," *Film Quarterly* 55, no. 3 (Spring 2002): 52–59.
20. Love Is Like Oxygen—Love Is a Many Splendored Thing—Up Where We Belong—All You Need Is Love—Lover's Game—I Was Made for Lovin' You—One More Night—Pride (in the Name of Love)—Don't Leave Me This Way—Silly Love Songs—Heroes—I Will Always Love You—Your Song.
21. See Neal Gershenfeld, *Fab: The Coming Revolution on Your Desktop—from Personal Computers to Personal Fabrication* (New York: Basic Books, 2007).
22. Christopher Noxon, *Rejuvenile: Kickball, Cartoons, Cupcakes, and the Reinvention of the American Grown-up* (New York: Crown, 2006).

23. This impulse is, of course, not new but resurfaces regularly. See Susan Tucker, *The Scrapbook in American Life* (Philadelphia: Temple University Press, 2006).
24. Joshua Driggs aka ZapWizard, "The Real Wood iPod," available at <http://www.macmod.com/content/view/363/101/>.
25. There has long been a split in the literature on gaming between narratologists, who emphasize the "stories" that video games generate, and the ludologists, who concentrate on game play as primary. See Ian Bogost, *Unit Operations: An Approach to Videogame Criticism* (Cambridge, MA: MIT Press, 2006), 68.
26. The sixth chapter of R. Buckminster Fuller's *Critical Path* (New York: St. Martin's Press, 1981) is devoted to the world game, and the Buckminster Fuller Institute maintains a page on its cite with numerous resources, available at <http://www.bfi.org/our_programs/who_is_buckminster_fuller/design_science/world_game>.

CHAPTER 4: WEB n.0

1. See <http://flickr.com>; <http://www.del.icio.us>. The term was coined by Thomas Vander Wal. See Daniel H. Pink, "Folksonomy," *New York Times Sunday Magazine*, December 11, 2005, available at <http://www.nytimes.com/2005/12/11/magazine/11ideas1-21.html?ex=1291957200&en=50937f27a0973e6e&ei=5090&partner=rssuserland&emc=rss>.
2. I developed many of these ideas in the course of writing the review "Welcome to Web 2.0: David Weinberger's *Everything Is Miscellaneous: The Power of the New Digital Disorder*," *Los Angeles Times Sunday Book Review*, June 17, 2006, R.11.
3. If there really is any remnant of progress left in the world of human affairs, the future will not understand this reference to a time when YouTube videos of cats using toilets attracted literally millions of human hours of viewing time.
4. This aphorism was one of my contributions to Mieke Gerritzen and Geert Lovink, *Mobile Minded* (Amsterdam: bis, 2002).
5. Jonathan Zittrain, *The Future of the Internet and How to Stop It* (New Haven, CT: Yale University Press, 2008).
6. Winston Churchill's maxim "Don't take No for an answer. Never submit to failure." has to be one of the most repeated pieces of advice to sales trainees everywhere. Quoted in Richard Langworth, *Churchill by Himself: The Definitive Collection of Quotations* (New York: Public Affairs, 2008), 569; originally in Winston Churchill, *My Early Life: A Roving Commission* (London: Thornton Butterworth, 1930), 74.
7. Jane Jacobs, *The Death and Life of Great American Cities* (1961; repr., New York: Modern Library, 1993; Jane Jacobs, *Systems of Survival: A Dialogue on the Moral Foundations of Commerce and Politics* (New York: Random House, 1992).

8. Manuel DeLanda offers a succinct definition of the difference between hierarchies and meshworks in "Homes: Meshwork or Hierarchy?" *Mediamatic*, available at <http://www.mediamatic.net/article-5914-en.html>. "Hierarchies are structures in which components have been sorted out into homogeneous groups, then articulated together. Meshworks, on the other hand, articulate heterogeneous components as such, without homogenizing." See also Manuel DeLanda, "Meshworks, Hierarchies, and Interfaces," in *The Virtual Dimension: Architecture, Representation, and Crash Culture*, ed. John Beckmann (New York: Princeton Architectural Press, 1998), 279–285.
9. See Boyce Rensberger. *Life Itself: Exploring the Realm of the Living Cell* (Oxford: Oxford University Press, 1998).
10. Miriam Hansen, "Of Mice and Ducks: Benjamin and Adorno on Disney," *South Atlantic Quarterly* 92, no. 1 (Winter 1993): 27–61.
11. See <http://wiki.creativecommons.org/History>; <http://wiki.creativecommons.org/Legal_Concepts#ip>
12. I am indebted to Lessig's work of over a decade defining, defending, and promoting Creative Commons, open-source culture, and the remix economy: *Remix: Making Art and Commerce Thrive in the Hybrid Economy* (New York: Penguin Press, 2008); *Code v2 and Other Laws of Cyberspace* (New York: Basic Books, 2007); *Free Culture: How Big Media Uses Technology and the Law to Lock Down Creativity* (New York: Penguin Press, 2004); *The Future of Ideas: The Fate of the Commons in a Connected World* (New York: Random House, 2001); *Code and Other Laws of Cyberspace* (New York: Basic Books, 1999).
13. Formerly at <http://www.jennyeverywhere.com>; now available at <http://theshifterarchive.com>.
14. This neologism is credited to journalist Jeff Howe in his article "The Rise of Crowdsourcing" *Wired* 14.06 (June 2006): 176–183. <http://www.wired.com/wired/archive/14.06/crowds.html>. See his book, *Crowdsourcing: Why the Power of the Crowd Is Driving the Future of Business* (New York: Crown Business, 2008) and blog, http://crowdsourcing.typepad.com/. His "white paper" version of the definition of crowdsourcing specifically foregrounds the economic relationships: "Crowdsourcing is the act of taking a job traditionally performed by a designated agent (usually an employee) and outsourcing it to an undefined, generally large group of people in the form of an open call."
15. James Joyce, *Ulysses* (1922; repr., Oxford: Oxford University Press, 1998), 182.
16. D. T. Max, "The Injustice Collector: Is James Joyce's Grandson Suppressing Scholarship?" *New Yorker*, June 19, 2006, 34-43. For an overview, see Robert Spoo, "Litigating the Right to Be a Scholar" *Joyce Studies Annual* 2008, 12–21, available at <http://muse.jhu.edu/journals/joyce_studies_annual/v2008/2008.spoo.html>

CHAPTER 5: BESPOKE FUTURES

1. Another way of describing this period is as the "long decade" of the 1990s, running from 1989 to 2001. I adopt this notion of long and short decades from the ways in which historians have proposed that the nineteenth was a long century, from the French Revolution in 1789 to the outbreak of World War I in 1914, and that the twentieth was a short one, running from 1914 to 1989. Likewise we can say that in the United States, the 1960s were a long decade, lasting from 1957 to 1973 (roughly the publication of Jack Kerouac's *On the Road* to the triple shocks of the OPEC oil embargo, Watergate, and the loss in Vietnam).
2. Francis Fukuyama, *The End of History and the Last Man* (New York: Free Press, 1992).
3. For a sterling analysis of New Economy hubris, see Thomas Frank, *One Market under God: Extreme Capitalism, Market Populism, and the End of Economic Democracy* (New York: Doubleday, 2000).
4. This figure comes from Lawrence Haverty Jr., senior vice president of State Street Research, quoted in Rachel Konrad, "Assessing the Carnage: Sizing Up the Market's Swift Demise," *CNET News*, March 8, 2001, available at <http://news.com.com/2009-1017-253125-2.html?legacy=cnet>.
5. In 1997, Bran Ferren of Disney Imagineering proclaimed, "The Net, I guarantee you, really is fire. I think it's more important than the invention of movable type." Quoted in Richard Rhodes, *Visions of Technology: A Century of Vital Debate about Machines, Systems, and the Human World* (New York: Simon and Schuster, 1999), 13.
6. Peter Lunenfeld, "TEOTWAWKI," *artext* 65 (1999): 34–35; reprinted in Peter Lunenfeld, *USER: InfoTechnoDemo* (Cambridge, MA: MIT Press, 2005).
7. See the concluding chapter, "Ublopia or Otivion," in Bruce Sterling, *Shaping Things*, design Lorraine Wild (Cambridge, MA: MIT Press, 2005), 138–145.
8. Bruce Mau, *Massive Change* (London: Phaidon, 2004), 15. There is an evident thirst in the design community for relevance and purpose. In 2000, *Adbusters* helped to commission "First Things First Manifesto." Signed by a host of luminaries including Irma Boom and Tibor Kalman, and inspired by Ken Garland's original from 1964, the manifesto was published simultaneously by a consortium of six of the major design magazines, including *Eye* and *Émigré*. The following quote gives a flavor of the language and intent. "There are pursuits more worthy of our problem-solving skills. Unprecedented environmental, social and cultural crises demand our attention. Many cultural interventions, social marketing campaigns, books, magazines, exhibitions, educational tools, television programs, films, charitable causes and other information design projects urgently require our expertise and help." Both the 2000 and 1964 "First Things First Manifesto" are available at <http://idie.enclavexquise.com/ftf2000-1964.html> (accessed March 30, 2005).

9. See Mieke Gerritzen and Geert Lovink, *Everyone Is a Designer: Manifest for the Design Economy* (Amsterdam: BIS Publishers, 2000) and their follow-up *Everyone Is a Designer in the Age of Social Media* ((Amsterdam: BIS Publishers, 2010).
10. Michael Beirut, "Warning: May Contain Non-Design Content," in *Seventy-nine Short Essays on Design* (New York: Princeton Architectural Press, 2007), 13.
11. Lauren G. Leighton, "The Great Soviet Debate over Romanticism: 1957-1964," *Studies in Romanticism*, v. 22, n. 1 (Spring, 1983): 41-64, 44.
12. Jamey Gambrell, "The Wonder of the Soviet World," *New York Review of Books* 41, no. 21 (December 22, 1994): 30-35.
13. I am here referring to H. G. Wells's classic science-fictional/futurist text *The Shape of Things to Come*, first published in 1929.
14. Karl Popper, *The Open Society and Its Enemies: Hegel and Marx* (1945; repr., London: Routledge, 2003), 262.
15. Gerard Jones, *Men of Tomorrow: Gangsters, Geeks, and the Birth of the Comic Book* (New York: Basic Books, 2004), 31–32.
16. William Gibson, "The Gernsback Continuum," in *Burning Chrome* (New York: Arbor House, 1986), 33.
17. See Peter Lunenfeld, "Bespoke Futures: Media Design and the Future of the Future," Adobe Design Center, Think Tank, June 19, 2007, available at <http://www.adobe.com/designcenter/thinktank/lunenfeld.html>.
18. Different fields use the term scenario to mean different things. Thus, the Global Business Network model is different from that used by researchers in human-computer interaction, which is not the same as those involved in software testing, which is in turn distinct from the way that design and market research use the word. For human-computer interaction, see, for example, John M. Carroll, *Making Use: Scenario-Based Design of Human-Computer Interactions* (Cambridge, MA: MIT Press, 2000); Mary Beth Rosson and John M. Carroll, *Usability Engineering: Scenario-Based Development of Human Computer Interaction* (Burlington, MA: Morgan Kaufmann, 2001). For the use of the term in software analysis, see J. Ryser and M. Glinz, "A Practical Approach to Validating and Testing Software Systems Using Scenarios," in *Proceedings QWE'99: Third International Software Quality Week Europe*, Brussels, November 1999, available at <http://www.ifi.unizh.ch/groups/req/ftp/papers/QWE99_ScenarioBasedTesting.pdf>; P. Hsia, J. Samuel, J. Gao, D. Kung, Y. Toyoshima, and C. Chen, "Formal Approach to Scenario Analysis," IEEE Software 11, no. 2 (1994):. 33–41. In marketing and design research, see Brenda Laurel, ed., *Design Research: Methods and Perspective* (Cambridge, MA: MIT Press, 2003).
19. Here, I refer to the controversial mentor of the futurist/scenario community Herman Kahn, author of the cold warrior's manual on scenario planning for nuclear conflict, *Thinking about the Unthinkable* (New York: Horizon, 1962).

20. Pierre Wack, "The Gentle Art of Reperceiving: Scenarios: Uncharted Waters Ahead," *Harvard Business Review* (September–October 1985): 73–89; Pierre Wack, "The Gentle Art of Reperceiving: Scenarios: Shooting the Rapids," *Harvard Business Review* (November–December 1985): 2–14; Kees van der Heijden, *Scenarios: The Art of Strategic Conversation* (Hoboken, NJ: John Wiley and Sons, 1996); James A. Ogilvy, *Creating Better Futures: Scenario Planning as a Tool for a Better Tomorrow* (Oxford: Oxford University Press, 2002); Peter Schwartz, *The Art of the Long View: Planning for the Future in an Uncertain World* (New York: Bantam Doubleday Dell, 1991–1996). Ogilvy and Schwartz cofounded the Global Business Network, an international consultancy that serves as an information clearinghouse for their favored methods of scenario planning; see <http://www.gbn.com/>.
21. Available at <http://www.gbn.com/ArticleDisplayServlet.srv?aid=34550>.
22. Schwartz, *The Art of the Long View*, p223
23. See <http://www.englishcut.com/archives/000004.html>.
24. See <http://www.vodafone.com/flash/futures/>; <https://motofuture.motorola.com/flash.html>.
25. See Geert Lovink, "Enemy of Nostalgia, Victim of the Present: Interview with Peter Lunenfeld," in *Uncanny Networks: Dialogues with the Virtual Intelligentsia* (Cambridge, MA: MIT Press, 2003).
26. The classic introduction to the study of chaos is James Gleick, *Chaos: Making a New Science* (New York: Penguin 1988). Two volumes linking chaos and cultural studies are N. Katherine Hayles, ed., *Chaos Bound: Orderly Disorder in Contemporary Literature and Science* (Ithaca, NY: Cornell University Press, 1990); N. Katherine Hayles, *Chaos and Order: Complex Dynamics in Literature and Science* (Chicago: University of Chicago Press, 1991).
27. A scan of "The Development of Abstract Art" is available at <http://www.art-history-online.info/imagepages/ahom03w07/ahom03w07barrdiagram.htm>. A downloadable movie of a Lorenz strange attractor is available at <http://hypertextbook.com/chaos/movies/lorenz.mov>.
28. Here I am caging the title of the third book of Neal Stephenson's Baroque Cycle. Stephenson was of course caging Isaac Newton. Neal Stephenson, *The System of the World* (New York: William Morrow, 2004).
29. In *The World Is Flat: A Brief History of the Twenty-first Century* (New York: Farrar, Straus and Giroux, 2005), neoliberal journalist Thomas L. Friedman looks at these same conditions and sees within them the seeds of what he calls Globalism 3.0.
30. Schwartz, *The Art of the Long View*, 12, 49.
31. My thanks to design researcher and Parsons professor Lisa Grocott for pushing me to emphasize the "designerly" aspect of bespoke futures.
32. MaSAI unintentionally references the Masai tribe of West African warriors, a bit of synchronicity that reflects Brian Eno's famous call for more Africa in computing: "What's pissing me off is that it uses so little of my body. You're

just sitting there, and it's quite boring. You've got this stupid little mouse that requires one hand, and your eyes. That's it. What about the rest of you? No African would stand for a computer like that. It's imprisoning." Available at <http://www.wired.com/wired/archive/3.05/eno_pr.html>.

33. From <http://stardustathome.ssl.berkeley.edu/background.html>:"In January 2004, the Stardust spacecraft flew through the coma of comet Wild2 and captured thousands of cometary dust grains in special aerogel collectors. Two years later, in January 2006, Stardust will return these dust grains—the first sample return from a solid solar-system body beyond the Moon—to Earth. But Stardust carries an equally important payload on the opposite side of the cometary collector: the first samples of contemporary interstellar dust ever collected. As well as being the first mission to return samples from a comet, Stardust is the first sample return mission from the Galaxy. But finding the incredibly tiny interstellar dust impacts in the Stardust Interstellar Dust Collector (SIDC) will be extremely difficult. We are seeking volunteers to help us to search for these tiny samples of matter from the galaxy. Volunteers are critical to the success of this project. Please help us find the first samples of contemporary Stardust ever collected."
34. Isotypes use an abstracted pictorial grammar "to create narrative visual material, avoiding details which do not improve the narrative character." Otto Neurath, *International Picture Language: The First Rules of Isotype* (London: K. Paul, Trench, Trubner and Company, 1936), 240. For more, see Karl H. Müller, "Otto Neurath and Contemporary Knowledge and Information Societies: A Newly Established Liaison," in *Encyclopedia and Utopia: The Life and Work of Otto Neurath (1882–1945)*, ed. Elisabeth Nemeth and Friedrich Stadler (Dordrecht: Kluwer, 1996), 135–142; George Pendle, "Otto Neurath's Universal Silhouettes," *Cabinet* 24 (Winter 2006–2007), available at <http://www.cabinetmagazine.org/issues/24/pendle.php>.
35. Frank Hartmann, "Humanization of Knowledge through the Eye," in *Making Things Public: Atmospheres of Democracy*, ed. Bruno Latour and Peter Weibel (Cambridge, MA: MIT Press, 2005), 707.
36. Brenda Laurel, *Utopian Entrepreneur* , design Denise Gonzales Crisp (Cambridge, MA: MIT Press, 2001), 69, 70.
37. See <http://manyeyes.alphaworks.ibm.com/manyeyes/page/About.html>. Many Eyes was developed by Martin Wattenberg and others in IBM's Collaborative User Research Experience Group starting in 2004.
38. William James, *Pragmatism* (1907; repr., Cambridge, MA: Harvard University Press, 1975), 137.
39. Though passivity, pessimism, and nihilism are hardly without their own partisans. See, most recently, Joshua Foa Dienstag, *Pessimism: Philosophy, Ethic, Spirit* (Princeton, NJ: Princeton University Press, 2006).
40. John Dewey, *Experience and Nature* (Chicago: Open Court, 1929), 45.

41. Unfortunately this list of cities and dates is not likely to end here.
42. The late Gould referred to them as NOMA, or "nonoverlapping magisteria," in an article of the same name, *Natural History* 106 (March 1997): 16–22, available at http://stephenjaygould.org/library/gould_noma.html
43. Quoted in Paulo Moura, "The Time of Killing," *Harper's Magazine* 309, no. 1850 (July 2004): 25. For an extended analysis, see Jon Ronson, *Them: Adventures with Extremists* (2001; repr., New York: Simon and Schuster, 2003).
44. Susan Sontag, *Against Interpretation and Other Essays* (New York: Farrar, Straus and Giroux, 1961).
45. Elizabeth L. Eisenstein, *The Printing Press as an Agent of Change* (Cambridge: Cambridge University Press, 1980).
46. Howard Gardner, *Frames of Mind: The Theory of Multiple Intelligences* (New York: Basic Books, 1983).
47. See Bob Stein, "We Could Be Better Ancestors Than This," in *The Digital Dialectic: New Essays on New Media*, ed. Peter Lunenfeld (Cambridge, MA: MIT Press, 1999).

GENERATIONS: HOW THE COMPUTER BECAME OUR CULTURE MACHINE

1. "We need a name for the new replicator, a noun that conveys the idea of cultural transmission, or a unit of *imitation*. 'Mimeme' comes from a suitable Greek root, but I want a monosyllable that sounds a bit like 'gene.' I hope my classicist friends will forgive me if I abbreviate mimeme to *meme*. If it is any consolation, it could alternatively be thought of as being related to 'memory,' or to the French word *même*. It should be pronounced to rhyme with 'cream.'" Richard Dawkins, *The Selfish Gene* (1976; repr., Oxford: Oxford University Press, 2006), 192.
2. Howard Rheingold, *Tools for Thought: The History and Future of Mind-Expanding Technology (*1985; repr., Cambridge, MA: MIT Press, 2000); available at <http://www.rheingold.com/texts/tft/>.
3. See Vannevar Bush, *Science: The Endless Frontier* (Washington, DC: U.S. Government Printing Office, 1945).
4. For an analysis of this transformation, see Paul N. Edwards, *Closed World: Computers and the Politics of Cold War America* (Cambridge, MA: MIT Press, 1996).
5. G. Pascal Zachary, *Endless Frontier: Vannevar Bush, Engineer of the American Century* (Cambridge, MA: MIT Press, 1999).
6. "The human mind . . . operates by association. With one item in its grasp, it snaps instantly to the next that is suggested by the association of thoughts, in accordance with some intricate web of trails carried by the cells of the brain. It has other characteristics, of course; trails that are not frequently followed

are prone to fade, items are not fully permanent, memory is transitory. Yet the speed of action, the intricacy of trails, the detail of mental pictures, is awe-inspiring beyond all else in nature." Vannevar Bush, "As We May Think," reprinted in Noah Wardrip-Fruin and Nick Montfort, eds., *The New Media Reader* (Cambridge, MA: MIT Press, 2003), 44. The essay is available many places online, including at <http://www.theatlantic.com/doc/194507/bush>.

7. Peter Galison, "War against the Center," *Grey Room* 4 (Summer 2001): 7–33.
8. M. Mitchell Waldrop, *The Dream Machine: J.C.R. Licklider and the Revolution That Made Computing Personal* (New York: Penguin, 2002).
9. Reprinted in Wardrip-Fruin and Montfort, *The New Media Reader*, 73–82.
10. AT&T, GE, and Philips have all been corporate entities since the late nineteenth century.
11. There are a huge number of books on IBM, and the company maintains an official history site, available at <http://www.ibm.com/ibm/history/>, but for the purposes here, the best source is a candid memoir, Thomas J. Watson and Peter Petre, *Father, Son, and Co.: My Life at IBM and Beyond* (New York: Bantam, 2000).
12. The United States has produced a staggering library of books about the art of selling. One of the most important of these from the first half of the twentieth century is Napoleon Hill's motivational best seller, *Think and Grow Rich*, first published in 1937, the title of which perfectly combines cognition and hucksterism.
13. It was while at Fairchild Semiconductor that Moore coined his *eponymous* law. As he wrote in "Cramming More Components onto Integrated Circuits," *Electronics Magazine*, April 19, 1965: "The complexity for minimum component costs has increased at a rate of roughly a factor of two per year. . . . Certainly over the short term this rate can be expected to continue, if not to increase. Over the longer term, the rate of increase is a bit more uncertain, although there is no reason to believe it will not remain nearly constant for at least 10 years. That means by 1975, the number of components per integrated circuit for minimum cost will be 65,000. I believe that such a large circuit can be built on a single wafer."
14. See Thierry Bardini, *Bootstrapping: Douglas Engelbart, Coevolution, and the Origins of Personal Computing* (Palo Alto, CA: Stanford University Press, 2001).
15. See John Markoff, *What the Dormouse Said: How the 60s Counterculture Shaped the Personal Computer* (New York: Viking, 2005). For a broader sense of the California spiritual landscape, see Erik Davis, *Visionary State* (San Francisco: Chronicle Books, 2006).
16. There is no book-length biography of Kay, although there should be. Many writings are available on the Web, and a good reminiscence is Alan Kay, "User Interface: A Personal View," in *The Art of Human Computer Interface Design*, ed. Brenda Laurel (Reading, MA: Addison-Wesley, 1990), 191–207.

17. Kay anticipated wireless networking as well. From a memo he wrote in 1971 to Xerox, available at <http://www.artmuseum.net/w2vr/archives/Kay/01_Dynabook.html>: "Though the Dynabook will have considerable local storage and will do most computing locally, it will spend a large percentage of its time hooked to various large, global information utilities which will permit communication with others of ideas, data, working models, as well as the daily chit-chat that organizations need in order to function. The communications link will be by private and public wires and by packet radio. Dynabooks will also be used as servers in the information utilities. They will have enough power to be entirely shaped by software."
18. Kay was especially impressed by the ways in which MIT mathematician Seymour Papert used Piaget's theories when he developed the LOGO programming language.
19. See Michael A. Hiltzik, *Dealers of Lightning: Xerox PARC and the Dawn of the Computer Age* (New York: HarperCollins, 1999); Douglas K. Smith and Robert C. Alexander, *Fumbling the Future: How Xerox Invented, Then Ignored, the First Personal Computer* (New York: William Morrow, 1988).
20. I take this phrase from Robert X. Cringely's documentary for the Public Broadcasting System, *The Triumph of the Nerds: The Rise of Accidental Empires* (1996), which drew from his earlier book *Accidental Empires: How the Boys of Silicon Valley Make Their Millions, Battle Foreign Competition, and Still Can't Get a Date* (New York: Harper Business, 1993).
21. Journalist Steve Levy took the title of his book, *Insanely Great: The Life and Times of Macintosh, the Computer That Changed Everything* (New York: Viking Penguin, 1993), from Jobs's own insistence that he and his team were on the verge of creating something not just functional or even profitable but rather insanely great. As for William Henry "Bill" Gates III, James Wallace's business book on his quest for dominance published the same year was titled *Hard Drive: Bill Gates and the Making of the Microsoft Empire* (New York: Harper Business, 1993).
22. See <http://www.w3.org/History/1989/proposal>.
23. The Swiss have four official languages, and use the linguistically neutral CH on their license plates and elsewhere to identify the country. In this Web page address, then, .ch is the DNS identifier for the country, as .uk is for England or .it is for Italy (that the United States does not use a DNS country code has been taken as imperial hubris by many).
24. Samir Chopra and Scott Dexter, *Decoding Liberation: A Philosophical Investigation of Free Software* (London: Routledge, 2007), 81. The full text of the jury's statement is archived on the Ars Electronica site, available at <http://90.146.8.18/en/archives/prix_archive/prix_projekt.asp?iProjectID=2183>.
25. See <http://www.gnu.org/gnu/thegnuproject.html>. See also Richard M. Stallman, *Free Software, Free Society: Selected Essays of Richard M. Stallman*

(Boston: Free Software Foundation, 2002); Sam Williams, *Free as in Freedom: Richard Stallman's Crusade for Free Software (*Boston: O'Reilly, 2002).

26. See Linus Torvalds and David Diamond, *Just for Fun: The Story of an Accidental Revolutionary* (New York: HarperCollins, 2001).
27. Computer and social scientist Paul Dourish was the first to tell me the joke that "Linux is free only if the value of your time is zero."
28. Eric S. Raymond, *The Cathedral and the Bazaar: Musings on Linux and Open Source by an Accidental Revolutionary* (Cambridge, MA: O'Reilly, 1999), available at <http://www.catb.org/˜esr/writings/cathedral-bazaar/>.
29. In the corporation's own words, from "Ten Things Google Has Found to Be True," available at <http://www.google.com/intl/en/corporate/tenthings.html>: PageRank™ "evaluates all of the sites linking to a web page and assigns them a value, based in part on the sites linking to them. By analyzing the full structure of the web, Google is able to determine which sites have been 'voted' the best sources of information by those most interested in the information they offer. This technique actually improves as the web gets bigger, as each new site is another point of information and another vote to be counted."
30. The advent of the more cluttered and text-rich iGoogle pages signal a shift in this approach. As of 2008, the upper menu allowed a user to toggle between iGoogle and the perhaps unintentionally funny "Classic Home," the stripped-down, original concept.

All hyperlinks current as of October 1, 2010

INDEX

www.secret-war.com